New Generation of Antipsychotic Drugs:
Novel Mechanisms of Action

International Academy for Biomedical and Drug Research

Vol. 4

Series Editors
S.Z. Langer, Paris
J. Mendlewicz, Brussels
G. Racagni, Milan

KARGER

Basel · Freiburg · Paris · London · New York · New Delhi · Bangkok · Singapore · Tokyo · Sydney

Workshop, Monte Carlo, March 16–18, 1992

New Generation of Antipsychotic Drugs: Novel Mechanisms of Action

Volume Editors
N. Brunello, Milan
J. Mendlewicz, Brussels
G. Racagni, Milan

21 figures and 27 tables, 1993

KARGER

Basel · Freiburg · Paris · London · New York · New Delhi · Bangkok · Singapore · Tokyo · Sydney

International Academy for Biomedical and Drug Research

Library of Congress Cataloging-in-Publication Data
New generation of antipsychotic drugs: novel mechanisms of action: workshop, Monte Carlo, March 16–18, 1992 / volume editors, N. Brunello, J. Mendlewicz, G. Racagni.
(International Academy for Biomedical and Drug Research; vol. 4)
Includes bibliographical references and index.
1. Antipsychotic drugs. I. Brunello, Nicoletta. II. Mendlewicz, J. III. Racagni, Giorgio.
IV. Series: International Academy for Biomedical and Drug Research (Series); vol. 4.
[DNLM: 1. Tranquilizing Agents, Major – pharmacology – congresses.]
ISBN 3–8055–5654–3

Drug Dosage

The authors and the publisher have exerted every effort to ensure that drug selection and dosage set forth in this text are in accord with current recommendations and practice at the time of publication. However, in view of ongoing research, changes in government regulations, and the constant flow of information relating to drug therapy and drug reactions, the reader is urged to check the package insert for each drug for any change in indications and dosage and for added warnings and precautions. This is particularly important when the recommended agent is a new and/or infrequently employed drug.

Printed in Switzerland on acid-free paper by Thür AG Offsetdruck, Pratteln
ISBN 3–8055–5654–3

Contents

Brunello N, Mendlewicz J, Racagni G (eds): New Generation of Antipsychotic Drugs: Novel Mechanisms of Action. Int Acad Biomed Drug Res. Basel, Karger, 1993, vol 4, pp 1–14

Multiple Dopamine Receptors as Molecular Targets for Antipsychotics

Jean-Charles Schwartz, Bruno Giros, Marie-Pascale Martres, Pierre Sokoloff

Unité de Neurobiologie et Pharmacologie (U.109) de l'INSERM, Centre Paul-Broca, Paris, France

Until recently, the existence of two dopamine receptor (DR) subtypes, D_1 and D_2, was generally accepted [1, 2] in spite of some proposals that additional subtypes may exist [3]. The use of molecular biology has dramatically modified this situation during the last two years and five receptor genes and several splice variants have been identified so far. D_2R was previously thought to represent the main target of antipsychotics but this hypothesis has to be reexamined in the light of recent findings.

Multiple DR Genes

The first DR gene cloned was that of D_2R [4]. The D_2R sequence paved the way for the cloning of a series of other DR genes based on their significant sequence homology. D_1R was the first to follow [5–7]. Then came the genes of a series of less expected receptors that markedly extend the DR family: the rat and human D_3 [8, 9], the human D_4 [10] and D_5 receptors [11], as well as the rat counterpart of D_5, termed D_{1B} [12]. The sequences of all these receptors reveal that they belong to a larger superfamily, that of receptors with seven transmembrane domains (7 TMs) and coupled to their intracellular transduction system by a GTP-binding (G) protein. The genes for both D_2R and D_4R are localized to human chromosome 11 [10, 13], D_1R to chromosome 5 [7], D_3R to chromosome 3 [14] and D_5R to chromosome 4 [12] (table 1).

Table 1. Synopsis of DR subtypes

	D_1	D_2	D_3	D_4	D_5
Coding sequence	446 A.A.	D_{2A} = 443 A.A. D_{2B} = 414 A.A.	400 A.A.	387 A.A.	477 A.A.
Chromosome localization	5 q31–q34	11 q22–q23	3 q13.3	11 p	4 p 16.3
Highest brain densities	Neostriatum	Neostriatum	Paleostriatum	Medulla Frontal cortex	Hypothalamus Hippocampus
Pituitary	No	Yes	No	?	No
DA neurons (A9, A10)	No	Yes	Yes	?	?
Affinity for DA	Micromolar	Micromolar	Nanomolar	Submicromolar	Submicromolar
Characteristic agonist	SKF-38393	Bromocriptine	7-OH-DPAT	?	SKF-38393
Characteristic antagonist	SCH-23390	Haloperidol	UH 232	Clozapine	SCH-23390
Adenylyl cyclase	Stimulates	Inhibits	?	Inhibits	Stimulates
Arachidonic acid release	(Potentiates)	Stimulates	?	?	?

A.A.: Amino acids.

Homologous Amino Acid Sequences of the DR Subtypes

All G-protein-coupled receptors comprise a pattern of 7 TMs of 20–25 amino acids connected by alternating extracellular and cytoplasmic loops composed of hydrophilic residues (fig. 1). The TMs form the ligand-binding domain, particularly three amino acid residues thought to interact with catecholamines: an aspartate in TM3 that forms an ion pair with the protonated amine group of dopamine (DA) and two serines in TM5 that presumably form a hydrogen bonding interaction with the two phenol groups of DA. The amino acid sequence of the human D_3R (fig. 1) displays 46% overall homology with that of D_2R, a value increasing to 78% when only the TMs are considered [9]. The overall homology between D_2R and D_4R is 41% [10]. The third cytoplasmic loop seems to be the main area for interaction with G proteins [15]. It is clear that DA receptors can be further subdivided by the length of this loop. Thus, D_1R and D_5R are characterized by a short third cytoplasmic loop and a long C-terminal tail and are coupled to Gs proteins. On the other hand D_2R, D_3R and D_4R, which display a long third cytoplasmic loop and a short C-terminal tail, might be coupled to Gi or Go proteins.

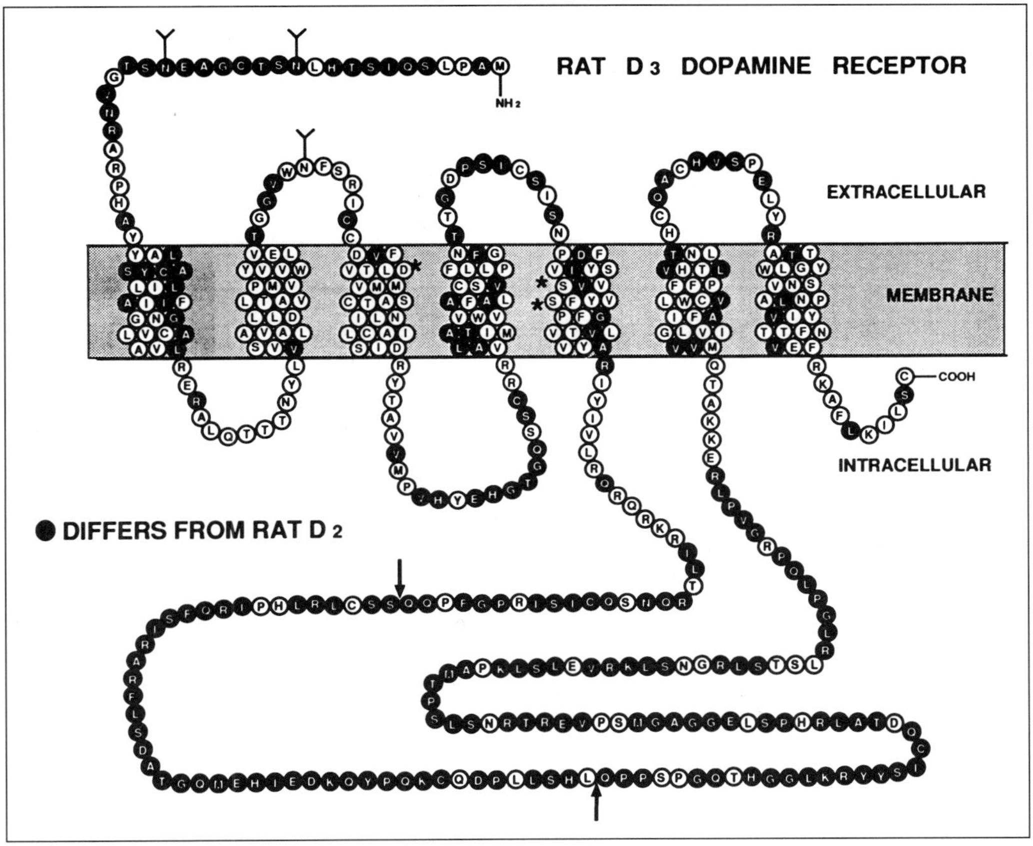

Fig. 1. Amino acid sequence of the rat D_3 receptor as compared to D_2R. Asterisks show amino acid residues presumably implicated in the binding of DA. The 46-amino-acid sequence between arrows lacks in the human D_3 receptor.

Alternative Splicing and Receptor Subtype Isoforms

The various genes of the DR family (fig. 2) can be classified in two groups according to their organization: (a) intronless genes, i.e. those of D_1R and D_5R, in which the coding nucleotide sequence is continuous; (b) genes with their coding sequence contained in discontinuous DNA segments (exons) interspersed among sequences (introns) that do not form a part of the mature mRNA. The latter organization, found in the D_2R, D_3R

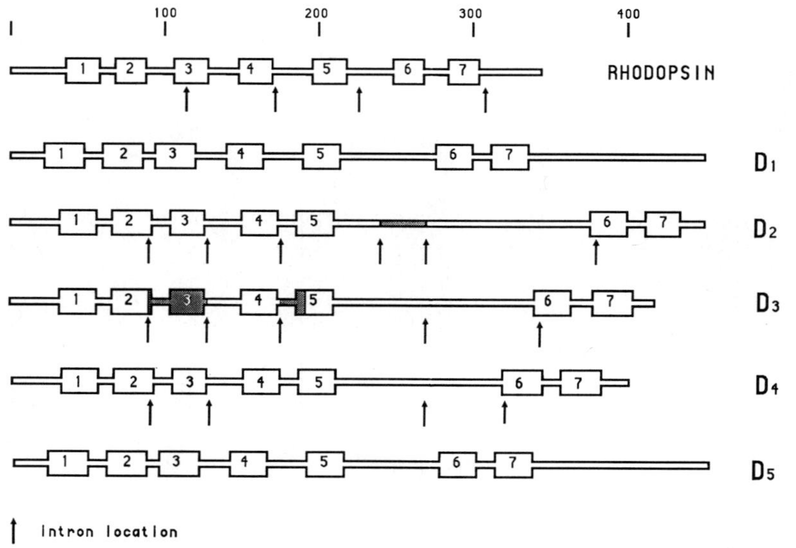

Fig. 2. The family of DR genes: organization and comparison with rhodopsin. The scale indicates the length of the amino acid sequence. Arrows indicate the position of introns; shaded areas correspond to alternative exons.

and D_4R genes, may potentially lead, through a mechanism of alternative splicing, to the biosynthesis of several distinct proteins encoded by a single gene.

For D_2R, alternative splicing potentially produces two receptors that differ by a stretch of 29 amino acids in the third cytoplasmic loop. These variants, or isoforms, are called $D_{2(444)}$ (or D_{2L} for D_2 long, or D_{2A}) and $D_{2(415)}$ (or D_{2S}, for D_2 short, or D_{2B}). They display identical pharmacology but have different patterns of expression among cerebral areas, although $D_{2(444)}$ predominates [16–18]. In addition, treatment with haloperidol, a D_2 antagonist, increases total D_2R mRNA in striatum and predominantly enhances the abundance of the smaller isoform in some other brain areas, e.g. the brainstem [19]. This presumably reflects increased receptor biosynthesis, a process which contributes to D_2R hypersensitivity and may be responsible for tardive dyskinesia.

Hence the two isoforms seem variously regulated and may differ at the second messenger level, through coupling to distinct G proteins, because such coupling probably involves the third cytoplasmic loop. Stimulation of both isoforms, however, inhibits adenylyl cyclase [17], activates K^+ channels and, via hydrolysis of membrane phospholipids, releases arachidonate, an intra- and extracellular second messenger [20, 21]. Nevertheless the two isoforms may differ slightly in the efficiency of their coupling with G proteins [16, 21, 22].

Two truncated forms of the D_3R mRNA, also generated by alternative splicing, have been detected in rat brain. In the first, a shift of the open reading frame results in the putative production of a protein of only 109 amino acids, unlikely to display receptor function. The second corresponds to a protein lacking an 18 amino acid sequence, still compatible with 7 TMs but no binding to such receptors was detected [23]. Hence the functional significance of the process of alternative splicing remains elusive, even more for D_3R than D_2R.

Intracellular Signalling Systems

Mammalian cell lines that do not normally express DRs, such as modified fibroblasts (e.g. CHO cells), can be transfected with cDNAs for the various DR subtypes. Then the pharmacology of these artificially expressed receptors and their signalling systems are studied. The obvious advantages of transfected cells is that they usually express a single receptor subtype at high density but one drawback is that they may differ from normal DA-responsive neurons in their complement of components of the intracellular signalling system, particularly the G proteins.

Using such tools, it has been confirmed that D_1R is positively coupled with adenylyl cyclase through the G protein known as Gs [5–7] and D_2R is negatively coupled to adenylyl cyclase through one of the Gi proteins [17, 24]. This agrees with the known properties of these receptors in DA-responsive neurons. These features suggest that the members of the subfamily of intronless receptors, i.e. D_1 and D_5, would be similarly coupled to Gs and those of receptors with introns, i.e. D_2, D_3 and D_4, to Gi proteins. In fact the pattern of adenylyl cyclase activation through D_5R stimulation is complex [25] and D_2R stimulation may also activate phospholipase C and increase intracellular Ca^{2+} in some transfected cells; in other cells, however, intracellular Ca^{2+} is decreased through activation of K^+ channels [26].

In addition, D_2R is also linked with phospholipase A_2. Both D_2R isoforms activate arachidonate release from transfected CHO cells, provided that phospholipase A_2 is simultaneously activated as a result of increased intracellular Ca^{2+} [20, 21]. This process, which is independent of inhibition of cAMP formation, is of potential functional importance because it occurs with low concentrations of D_2R agonists and arachidonic acid is a precursor for a variety of biologically active eicosanoids presumably involved in neurotransmission [27]. This signalling pathway may also account for the synergistic interaction between D_1R and D_2R: in CHO cells cotransfected with both subtypes, D_1R stimulation or application of a cAMP analog markedly enhances the release of arachidonate induced by D_2R agonists [21]. As arachidonate and its cytochrome P450 metabolites are potent inhibitors of (Na^+ + K^+) ATPase in neurons of the striatum, their formation may well account for the D_1/D_2 synergistic responses in vivo, which may involve this transporter [28].

There is no evidence for coupling of D_3R expressed in CHO cells, as judged from either guanyl-nucleotide-induced modulations of DA binding, inhibition of adenylyl cyclase or activation of arachidonate release [8, 29]. Hence the signalling system of D_3R remains to be identified but it seems likely that it couples optimally to (a) G proteins(s) distinct from those to which couples D_2R. The binding of DA at D_4R is sensitive to guanyl nucleotides, indicating effective coupling to endogenous G proteins in COS-7 cells but functional activity had not been reported [10]. In permanently transfected CHO cells D_4R, like D_2R, mediates both inhibition of cAMP formation and stimulation of arachidonate release.

Two major conclusions arise from these studies: (1) DA may affect target cells through the three major signalling pathways i.e. cAMP, phospholipase A_2 and phospholipase C products, and (2) the nature of responses to DA depends not only on the receptor subtype involved but also on the cell type in which it is coexpressed with a given set of G proteins and effector systems.

Pharmacology of DR Subtypes: The Agonists

The D_1-like and D_2-like subfamilies of DRs, distinguished on genetic and molecular grounds, can also be differentiated pharmacologically. Thus the 'D_1-selective' agonist SKF 38393 and antagonist SCH-23390 are almost equipotent at D_1R and D_5R, but 10^2–10^3 times less potent at 'D_2-

Table 2. Pharmacology of DR subtypes

	K_i values, n*M*				
	D_1	D_2	D_3	D_4	D_5
Agonists					
Dopamine[1]	2,000	2,000	30	450	250
Apomorphine	700	70	70	(4)[2]	400
Bromocriptine	700	5	7	300	500
Pergolide	1,400	20	2	–	900
Quinpirole	14,000	1,400	40	50	–
SKF 38393	150	10,000	5,000	10,000	100
Antagonists					
Haloperidol	30	0.6	3	5	40
Pimozide	–	10	11	40	–
Thioproperazine	–	0.5	1	50	–
(–)Sulpiride	40,000	10	20	50	80,000
Amisulpride	–	1.3	2.4	–	–
Carpipramine	–	8.7	15	–	–
Pipotiazine	–	0.2	0.3	–	–
UH 232	–	40	10	–	–
Clozapine	140	70	500	9	250
SCH-23390	0.3	1,000	1,000	3,500	0.3

[1] Value + presence of a guanyl nucleotide.
[2] High affinity component.

like' receptors; conversely, sulpiride is approximately 1,000-fold more potent at D_2R, D_3R or D_4R than at D_1R or D_5R (table 2). Within each subfamily the various subtypes differ in their apparent affinity for DA, which binds with micromolar K_i values at D_1R and D_2R but with values 5–10 times lower at D_4R and D_5R [10–12] and nearly 100-fold lower at D_3R [8, 29]. The unusually high affinity of DA for D_3R is rather intriguing but it may reflect the ability of the neurotransmitter to act at larger distance or to remain bound for longer periods than at other receptors.

Among DA agonists, only apomorphine displays similar affinities at all receptor subtypes; other agonists seem relatively specific for only a subset of receptors e.g. quinpirole at D_3 and D_4 or bromocriptine at D_2 and D_3 receptors. This selectivity profile is likely to have therapeutic implications regarding the efficacy and side effects for these agents when used to treat Parkinson's disease. In addition 7-OH-DPAT, an aminotetralin derivative, was recently identified as a highly selective D_3R agonist [30].

Pharmacology of DR Subtypes: Antipsychotics

Until recently, it has been generally assumed that neuroleptics derive their antipsychotic activity from D_2R blockade. Nevertheless, the discovery of several D_2-like receptors, well recognized by these drugs, raises the question: what are the cerebral targets for this antipsychotic activity? Using positron emission tomography, occupancy of D_2R in the brain of human subjects receiving therapeutic doses of various antipsychotics is between 40 and 100% [31]. As most antipsychotics display similar affinities at D_2R and D_3R (table 2), it seems likely that significant occupancy of D_3R also occurs in these subjects. This applies even in the case of haloperidol, which displays a rather high preference for D_2R and, therefore, it cannot be excluded that the antipsychotic activity is related to occupancy of D_3R, a hypothesis consistent with their distribution.

This may not hold true in all cases for D_4R, for which several antipsychotics, e.g. thioproperazine, have about 100-fold lower affinity. In addition, the low or even very low affinity of many antipsychotics, particularly the benzamide derivatives such as sulpiride, at D_1R and D_5R exclude the latter as targets for antipsychotics. Hence D_2R and D_3R remain the most likely common targets for most antipsychotics.

Nevertheless, differential blockade of D_2R, D_3R and D_4R may account for the therapeutic profile of some antipsychotics. As clozapine displays significantly higher affinity at D_4R than at D_2R or, even more, D_3R, its 'atypical' properties may derive from a rather selective blockade of D_4R (or both D_2R and D_4R). On the other hand, amisulpride, carpipramine, pipotiazine and pimozide, which belong to a group of antipsychotics with a 'disinhibitory' therapeutic profile, is almost as potent at D_3R than at D_2R and much less potent at D_4R (table 2).

The only antagonists with higher affinity at D_3R than D_2R are AJ 76 and UH 232, formerly identified in animal studies as 'autoreceptor selective' agents, i.e. with behavioral stimulant properties somewhat similar to those of DA agonists [32]. It will be of interest to assess the antipsychotic potency of such agents.

Localization of DR Subtypes

The D_1R and D_2R are the most abundantly expressed in brain, the corresponding mRNAs being detected easily in all main dopaminoceptive areas where these receptors were previously detected by autoradiography [33–35]. D_2R mRNA is detected by in situ hybridization [8, 36–38] in most areas where D_2R binding sites occur but with some exceptions. For instance, in the islands of Calleja or cerebellar lobules 9 and 10, [^{125}I]iodosulpride binding sites are found without D_2R mRNA: there, these sites presumably correspond to D_3R rather than D_2R since they colocalize with the D_3R mRNA [36]. The D_2R gene is expressed in all medium-sized enkephalin neurons of the caudate-putamen, nucleus accumbens, olfactory tubercle and piriform cortex, and large, possibly cholinergic, striatal neurons. A fraction of striatal neurons seem to coexpress D_1R and D_2R [38]. The two isoforms of D_2R seem to be coexpressed in most dopaminoceptive areas of the brain or pituitary, although in various ratios [16].

The expression of the D_3R gene is mainly restricted to some areas receiving DA inputs from the A_{10} cell group, e.g. the anterior part of nucleus accumbens, islands of Calleja, bed nucleus of the stria terminalis and other limbic areas such as septal or mammillary nuclei; the D_3R gene is also highly expressed in archicerebellum [36]. In most of these areas, the distributions of D_2R and D_3R mRNAs seem not to overlap, indicating that distinct cell populations are involved. Recently, with the design of [^{3}H]7-OH-DPAT as a selective D_3R probe, it was established that, in rat brain, the overall abundance of the D_3R protein is about two orders of magnitude lower than that of the D_2R protein; in addition its regional distribution, grossly parallel to that of D_3R mRNAs, suggests that it is mainly expressed in short neurons [30]. Hence the almost exclusive localization of D_3R to limbic areas thought to be involved in the control of emotional, endocrine and cognitive processes reinforces the idea that it may mediate (participate in) the antipsychotic action of neuroleptics. The D_3R message is almost undetectable in pituitary [8, 36]. D_4R mRNA distribution has only been established in a preliminary way in few brain areas; in human brain it seems more abundant in limbic than in striatal areas [10]. The distribution of D_5R mRNA [12] seems different from that of D_1R mRNA, being restricted mainly to lateral mammillary nuclei, anterior pretectal nuclei and some hippocampal layers.

Diverse Subtypes of DA Autoreceptors

Autoreceptors are a functional class of receptors expressed by DA neurons and mediating inhibitions of these neurons activity or of DA synthesis and release. These autoreceptors had been identified pharmacologically as D_2Rs. Accordingly, DA neurons in the substantia nigra pars compacta and ventral tegmental area express D_2 mRNA [36–38], as confirmed by 6-hydroxydopamine lesions [8]. Both $D_{2(444)}$ and $D_{2(415)}$ isoforms are expressed in rat substantia nigra [16, 19]. In addition, in substantia nigra pars compacta, particularly the lateral part, and the ventral tegmental area, DA neurons express D_3R mRNA, indicating that the D_3R is also an autoreceptor [8, 36]. This accounts for the pharmacology of D_3R for which agents previously coined 'autoreceptor-selective' agonists and antagonists show preference over D_2R. Other DA neurons (A_{13}, A_{14}, A_{16}) may also be endowed with D_3 autoreceptors [36]. In view of this diversity it has to be established what is, among the various roles attributed to autoreceptors in the control of DA neuron activity, the role of each receptor subtype (or isoform). It would also be important to establish whether a single DA neuron expresses several DR subtypes or isoforms.

Conclusion: DR Diversity and Antipsychotic Activity

The recent discovery of so many different DR subtypes produced by distinct genes or through alternative splicing raises questions about the mode of action of antipsychotics. Previously, it was generally admitted that cerebral D_2R was the common target of this class of compounds but the in vitro pharmacology of the two D_2R isoforms, D_3R and, to a large extent, D_4R suggests that these distinct proteins are significantly occupied in vivo during treatments by antipsychotics. Hence it remains to be established whether occupancy of a single DR subtype is sufficient to produce antipsychotic effects. These subtypes markedly differ in their distribution and occupancy of the widely expressed D_2R in many brain areas may account for the psychiatric effects as well as for the extrapyramidal and endocrine side effects occurring, although to a variable extent, with all available antipsychotics. Blockade of a significant fraction of D_3R can be expected under therapeutic conditions during which extensive D_2R occupancy was directly demonstrated using the PET scan methodology.

Selective occupancy of D_3R by compounds to be designed should be expected to result in blockade of DA transmissions in only few brain areas: using [^{3}H]7-OH-DPAT, the overall abundance of D_3R in brain was found to be two orders of magnitude lower than that of D_2R [30]. The absence of D_3R mRNA (and extremely low levels of [^{3}H]7-OH-DPAT binding sites) in striatum and pituitary suggests that selective D_3R antagonists should be devoid of the extrapyramidal and prolactin-enhancing side effects common to currently available antipsychotics. But will these compounds retain some antipsychotic activity? This remains to be established in spite of the occurrence of D_3R in limbic areas possibly involved in the emotional and cognitive disturbances of schizophrenia.

The case of D_4R is even less clear because its abundance and precise distribution in brain remain to be established and its pharmacology (in transiently transfected cells) suggests that it is only weakly blocked by several antipsychotics during therapy [10].

The design of highly selective D_3R or D_4R antagonists, which should be facilitated by the use of transfected cell lines expressing a single DR in high levels, should make it possible to verify these various hypotheses. In any case, the discrete expression of these subtypes in few dopaminoceptive areas of the brain offers unexpected opportunities to refine the 'dopamine hypothesis' of schizophrenia and, possibly, to develop novel and improved therapeutic strategies in this field.

References

1 Spano PF, Govoni S, Trabucchi M: Studies on the pharmacological properties of dopamine receptors in various areas of the central nervous system. Adv Biochem Psychopharmacol 1978;19:155–165.

2 Kebabian JW, Calne DB: Multiple receptors for dopamine. Nature 1979;277:93–96.

3 Schwartz JC, Delandre M, Martres MP, Sokoloff P, Protais P, Vasse M, Costentin J, Laibe P, Wermuth CG, Gulat C, Lafitte A: Biochemical and behavioral identification of discriminant benzamide derivatives: New tools to differentiate subclasses of dopamine receptors; in Usdin E, Carlsson A, Dahlström A, Engel J (eds): Catecholamines: Neuropharmacology and Central Nervous System – Theoretical Aspects. New York, Liss, 1984, pp 59–72.

4 Bunzow JR, Van Tol HHM, Grandy DK, Albert P, Salon J, Christie McD, Machida CA, Neve KA, Civelli O: Cloning and expression of a rat D_2 dopamine receptor cDNA. Nature 1988;336:783–787.

5 Zhou QZ, Grandy DK, Thambi L, Kushner JA, Van Tol HHM, Cone R, Pribnow D, Salon J, Bunzow JR, Civelli O: Cloning and expression of human and rat D_1 dopamine receptors. Nature 1990;347:76–80.

6 Dearry A, Gingrich JA, Falardeau P, Fremeau RT, Bates MD, Caron MG: Molecular cloning and expression of the gene for a human D_1 dopamine receptor. Nature 1990;347:72–76.

7 Sunahara RK, Niznik HB, Weiner DM, Stormann TM, Brann MR, Kennedy JL, Gelernter JE, Rozmahel R, Yang Y, Israel Y, Seeman P, O'Dowd BF: Human dopamine D_1 receptor encoded by an intronless gene on chromosome 5. Nature 1990; 347:80–83.

8 Sokoloff P, Giros B, Martres MP, Bouthenet ML, Schwartz JC: Molecular cloning and characterization of a novel dopamine receptor (D_3) as a target for neuroleptics. Nature 1990;347:146–151.

9 Giros B, Martres MP, Sokoloff P, Schwartz JC: cDNA cloning of the human dopaminergic D_3 receptor and chromosome identification. CR Séances Acad Sci 1990; 311:501–508.

10 Van Tol HHM, Bunzow JR, Guan HC, Sunahara RK, Seeman P, Niznik HB, Civelli O: Cloning of the gene for a human dopamine D_4 receptor with high affinity for the antipsychotic clozapine. Nature 1991;350:610–614.

11 Sunahara RK, Guan HC, O'Dowd BF, Seeman P, Laurier LG, Ng G, George SR, Torchia J, Van Tol HHM, Niznik HB: Cloning of the gene for a human dopamine D_5 receptor with higher affinity for dopamine than D_1. Nature 1991;350:614–619.

12 Tiberi M, Jarvie KR, Silvia C, Falardeau P, Gingrich JA, Godinot N, Bertrand L, Yang-Feng TL, Fremeau RT Jr, Caron MG: Cloning, molecular characterization, and chromosomal assignment of a gene encoding a second D_1 dopamine receptor subtype: Differential expression pattern in rat brain compared with the D_{1A} receptor. Proc Natl Acad Sci USA 1991;88:7491–7495.

13 Grandy DK, Marchionni MA, Makan H, Stofko RE, Alfano M, Prothingham L, Fischer JB, Burke-Howie KJ, Bunzow JR, Server AC, Civelli O: Cloning of the cDNA and gene for a human D_2 dopamine receptor. Proc Natl Acad Sci USA 1989; 84:9762–9766.

14 Le Coniat M, Sokoloff P, Hillion J, Martres MP, Giros B, Pilon C, Schwartz JC, Berger R: Chromosomal localization of the human D_3 dopamine receptor gene. Hum Genet 1991;87:618–620.

15 Strader DC, Sigal SI, Dixon AFR: Mapping of functional domains of the β-adrenergic receptor. Am J Respir Cell Mol Biol 1989;1:81–86.

16 Giros B, Sokoloff P, Martres MP, Riou JF, Emorine LJ, Schwartz JC: Alternative splicing directs the expression of two D_2 dopamine receptor isoforms. Nature 1989; 342:923–926.

17 Dal Toso R, Sommer B, Ewert M, Herb A, Pritchett DB, Bach A, Shivers BD, Seeburg PH: The dopamine D_2 receptor: Two molecular forms generated by alternative splicing. EMBO J 1989;8:4025–4034.

18 Monsma FJ, McVittie LD, Gerfen CR, Mahan LC, Sibley DR: Multiple D_2 dopamine receptors produced by alternative RNA splicing. Nature 1989;342:926–929.

19 Martres MP, Sokoloff P, Giros B, Schwartz JC: Effects of dopaminergic transmission interruption on the D_2 receptor isoforms in various cerebral tissues. J Neurochem 1992;58:673–679.
20 Kanterman RY, Mahan LC, Briley EM, Monsma FJ, Sibley DR, Axelrod J, Feldek CC: Transfected D_2 dopamine receptors mediate the potentiation of arachidonic release in Chinese hamster ovary cells. Mol Pharmacol 1991;39:364–369.
21 Piomelli D, Pilon C, Giros B, Sokoloff P, Martres MP, Schwartz JC: Dopamine activation of the arachidonic acid cascade via a modulatory mechanism as a basis for D_1/D_2 receptor synergism. Nature 1991;353:164–167.
22 Montmayeur JP, Borelli E: Transcription mediated by a cAMP-responsive promoter element is reduced upon activation of dopamine D_2 receptors. Proc Natl Acad Sci USA 1991;88:3135–3139.
23 Giros B, Martres MP, Pilon C, Sokoloff P, Schwartz JC: Shorter variants of the D_3 dopamine receptor produced through various patterns of alternative splicing. Biochem Biophys Res Commun 1991;176:1584–1592.
24 Neve KA, Henningen RA, Bunzow JR, Civelli O: Functional characterization of a rat dopamine D_2 cDNA expressed in a mammalian cell line. Mol Pharmacol 1989; 36:446–451.
25 Grandy DK, Zhang Y, Bouvier C, Zhou Q, Johnson RA, Allen L, Buck K, Bunzow JR, Salon J, Civelli O: Multiple human D_5 dopamine receptor genes: a functional receptor and two pseudogenes. Proc Natl Acad Sci USA 1991;88:9175–9179.
26 Vallar L, Muca C, Magni M, Albert P, Bunzow J, Meldolesi J, Civelli O: Differential coupling of dopaminergic D_2 receptors expressed in different cell types. J Biol Chem 1990;265:10320–10326.
27 Piomelli D, Greengard P: Lipoxygenase metabolites of arachidonic acid in neuronal transmembrane signalling. Trends Pharmacol Sci 1990;11:367–373.
28 Bertorello AM, Hopfield JE, Aperia A, Greengard P: Inhibition by dopamine of $(Na^+{+}K^+)$ ATPase activity in neostriatal neurons through D_1 and D_2 dopamine receptor synergism. Nature 1990;347:386–388.
29 Sokoloff P, Andrieux M, Besançon R, Pilon C, Martres MP, Giros B, Schwartz JC: Pharmacology of human D_3 dopamine receptor expressed in a mammalian cell line: Comparison with D_2 receptor. Eur J Pharmacol Mol Biol, in press.
30 Levesque D, Diaz J, Pilon C, Martres MP, Giros B, Souil E, Schott D, Morgat JL, Schwartz JC, Sokokoff P: Identification, characterization and localization of the dopamine D_3 receptor in rat brain using $[^3H]$7-hydroxy dipropylaminotetralin. Proc Natl Acad Sci USA (in press).
31 Farde L, Wiesel FA, Nordström AL, Sedvall G: D_1 and D_2 dopamine receptor occupancy during treatment with conventional and atypical neuroleptics. Psychopharmacology 1989;99:S28–S31.
32 Svensson K, Johansson AM, Magnusson T, Carlsson A: (+)-AJ 76 and (+)-UH 232: Central stimulants acting as preferential dopamine autoreceptor antagonists. Naunyn Schmiedebergs Arch Pharmacol 1986;334:234–245.
33 Dawson TM, Barone P, Sidhu A, Wamsley JK, Chase TN: The D_1 dopamine receptor in the rat brain: quantitative autoradiographic localization using an iodinated ligand. Neuroscience 1988;26:83–100.
34 Bouthenet ML, Martres MP, Sales N, Schwartz JC: A detailed mapping of dopamine D_2 receptors in rat central nervous system by autoradiography with $[^{125}I]$iodosulpride. Neuroscience 1987;20:117–155.

35 Fremeau RT, Duncan GE, Fornaretto MG, Dearry A, Gingrich JA, Brees GR, Caron MG: Localisation of D_1 dopamine receptor mRNA in brain supports a role in cognitive, affective, and neuroendocrine aspects of dopaminergic neurotransmission. Proc Natl Acad Sci USA 1991;88:3772–3776.

36 Bouthenet ML, Souil E, Martres MP, Sokoloff P, Giros B, Schwartz JC: Localization of dopamine D_3 receptor mRNA in the rat brain using in situ hybridization histochemistry: Comparison with dopamine D_2 receptor mRNA. Brain Res 1991;564: 203–219.

37 Le Moine C., Normand E, Guitteny AF, Rouque B, Teoule R, Block B: Dopamine receptor gene expression by enkephalin neurons in rat forebrain. Proc Natl Acad Sci USA 1990;87:230–234.

38 Mansour A, Meador-Woodruff JH, Bunzow JR, Civelli O, Akil H, Watson SJ: Localization of dopamine D_2 receptor mRNA and D_1 and D_2 receptor binding in the rat brain and pituitary. An in situ hybridization receptor autoradiographic analysis. J Neurosci 1990;10:2587–2601.

Jean-Charles Schwartz,
Unité de Neurobiologie et Pharmacologie (U.109) de l'INSERM,
Centre Paul-Broca, 2ter, rue d'Alésia, F–75014 Paris (France)

Brunello N, Mendlewicz J, Racagni G (eds): New Generation of Antipsychotic Drugs: Novel Mechanisms of Action. Int Acad Biomed Drug Res. Basel, Karger, 1993, vol 4, pp 15–26

Nondopaminergic Mechanisms Relevant to Antipsychotic Activity[1]

Zachary N. Stowe, Charles B. Nemeroff

Department of Psychiatry and Behavioral Sciences, Emory University School of Medicine, Atlanta, Ga., USA

Introduction

In the over 30 years since the introduction of antipsychotic drugs to the treatment modalities available for patients with schizophrenia, numerous investigators have sought to elucidate the mechanisms of the therapeutic actions of these agents. Initial receptor binding studies demonstrated a strong correlation between clinical potency and antagonism of the D_2-receptor [1, 2]. In contrast, there is poor correlation between clinical efficacy and the binding to other receptors [3]. However, the advent of clinically efficacious atypical antipsychotic drugs (e.g. clozapine, remoxipride) and the development of newer putative antipsychotics (e.g. BMY 14802, CI-943) with relatively low affinities for DA D_2-receptors have prompted the development of more integrated theories involving multiple neuronal systems. Recent pharmacological advances have implicated a potential role for several of the newly discovered DA receptors (D_3, D_4 and D_5) as well as the D_1 DA receptor in addition to non-DA receptors including serotonergic (5HT), noradrenergic, cholinergic, and sigma receptors as having a role in both the therapeutic action and side effect profiles of these agents [4 for review]. Despite the sundry of effects associated with antipsychotic drug treatment mediated by various receptor subtypes within the central nervous systems (CNS), the majority of the available data for both typical and atypical antipsychotic drugs continues to implicate a role for

[1] Supported by NIMH MH39415 (C.B.N.) and a Daland Fellowship from the American Philosophical Society (Z.N.S.).

dopamine (DA) neurotransmission modulation in their mechanism of action. A neuropeptide that has been shown to conclusively interact with and modulate CNS DA systems in neurotensin (NT). The concatenation of available preclinical and clinical data on NT makes it an attractive neurotransmitter candidate to mediate the action of antipsychotic drugs. In this monograph we will provide an overview of NT-DA interactions with an emphasis on recent developments.

Background

NT is an endogenous tridecapeptide (pGlu-Leu-Tyr-Glu-Asn-Lys-Pro-Arg-Arg-Pro-Tyr-Ile-Leu-OH) found both in the periphery [5 for review] and within the CNS [6, 7 for review]. NT has fulfilled virtually all of the requisite criteria for classification as a neurotransmitter. Both the genes encoding the NT precursor [8] and more recently the NT receptor [9] have been structurally characterized. Numerous preclinical investigations describing the anatomical, neurochemical, behavioral, electrophysiological, and pharmacological characteristics of NT continue to underscore its intimate relationship with CNS DA systems. NT appears to selectively modulate the mesolimbic and mesocortical dopaminergic systems via NT binding sites. However, the mechanism(s) by which NT exerts these modulatory effects remains obscure. The primary impetus behind the burgeoning number of studies seeking to elucidate the role(s) of NT in the CNS is the remarkable pharmacobehavioral similarities between NT and atypical antipsychotic drugs.

Neuroanatomy

NT displays a heterogeneous distribution through the CNS [10] and is closely associated with dopamine (DA) neurons in the mesencephalon, medial prefrontal cortex (MPF), and hypothalamus. Despite the proximal relationship of NT to CNS DA systems, the concentration of NT appears to be relatively independent from the loss of DA. The depletion of greater than 95% of DA from the ventral tegmental area (VTA), substantia nigra (SN), nucleus accumbens (NA), caudate nucleus, olfactory tubercles, and frontal cortex using 6-OHDA does not alter NT concentrations in these regions. Also no significant alterations in the regional concentration of NT is observed in postmortem brain tissue of patients with Parkinson's disease [11], while the density of NT receptors in the SN was decreased.

NT is particularly closely associated with the mesolimbic, nigrostriatal and mesocortical DA systems [7 for review]. Briefly, high concentrations of NT are found in close proximity to both DA neuronal perikarya [12] and DA terminal regions including the SN and VTA, the NA, and the caudate.

NT is colocalized within a subset of DA perikarya in the A8, A9, and A10 neurons in the mesencephalon [13, 14]. The concentrations of NT and DA are positively correlated in the mesolimbic projections and negatively correlated in the mesostriatal projections [15]. Virtually all of the MPF DA terminals in the rat also contain NT and some of these also contain cholecystokinin (CCK) [16, 17]. In contrast to the rat, NT and DA do not appear to be colocalized in humans. This may explain the lack of alteration in NT concentrations in the postmortem brain tissue of patients with Parkinson's disease noted above. NT receptors are found on DA cell bodies and certain DA terminals in the nigrostriatal system [18, 19]. In contrast to the mesencephalon, the density of NT receptors in the cerebral cortex is relatively high whereas endogenous NT concentrations are low. The distribution of NT receptors closely resembles the pattern of DA D_1-receptors [20, 21]. Several NT-containing projections have been identified in the rat bain [7 for review].

Neurochemistry

Central injection of NT, like parentally applied antipsychotic drugs, increases DA turnover in the NA, olfactory tubercles, and the diagonal band of Broca, as measured by the concentrations of the major dopamine metabolites, homovanillic acid and dihydroxyphenylacetic acid. In contrast, DA applied to the VTA causes autoreceptor activation and decreases turnover in these terminal regions [7]. Central application of NT also increases tyrosine hydroxylase activity [22] and stimulates DA release in the striatum [23], and NA [24]. NT application produces a greater increase in the metabolism of DA in the NA than in the striatum, and higher doses of NT partially reversed amphetamine-induced alterations in DA metabolism. NT does not alter basal DA-stimulated adenylate cyclase activity [25], and even high concentrations of NT do not affect [^{3}H]spiperone binding [25], suggesting that NT does not appear to function simply as a D_1

(adenylate cyclase-linked) or D_2 (spiperone labelled) receptor antagonist. However, one group has reported that NT reduces the binding affinity of [^{3}H]-n-propylnorapomorphine, a DA agonist, in the striatum and nucleus accumbens [26, 27]. The recent elucidation of the existence and structure of D_3, D_4, and D_5 DA receptor subtypes raises the question of whether NT may interact with one of these receptors, or alter the binding of DA to them.

Behavioral

Behavioral studies have shown that the direct injection of NT into the CNS, like systemically administered antipsychotic drugs, produces hypothermia [28], potentiation of barbiturate-induced sedation [29], catalepsy, blockade of psychostimulant-induced locomotor activity [30] and inhibition of avoidance, but not escape, responding in a discrete trial, conditioned avoidance paradigm [31]. Microinjection of NT into the DA, like parenterally applied antipsychotic drugs, blocks the increased locomotion and rearing produced by the indirect DA agonist methylphenidate, cocaine and *d*-amphetamine [30]. In contrast, NT does not alter *d*-amphetamine-induced stereotypy, an effect mediated by the nigrostriatal DA projection, underscoring the selective effects of NT on the mesolimbic DA system. Intraaccumbens injection of NT blocks the rewarding effects of electrical self-stimulation observed in rats with electrodes placed in the VTA [31]. Rodents will preferentially lever press for intra-VTA injection of NT while ignoring the inactive lever [32]. Thus, it has been proposed that NT may have a role in CNS reward mechanisms, by modulating DA neuronal activity [32].

Electrophysiological

The available electrophysiological literature continues to show similarity to clinically efficacious antipsychotic drugs [33 for review]. Briefly, NT has consistently been reported to produce an increase in the firing rates of DA neurons of the SN and VTA both in vivo [34, 35] and in vitro [36], while no effect on nondopaminergic neurons has been observed [42]. These excitations appear to be dose-dependent and occur via a calcium-dependent postsynaptic mechanism [36]. Direct application of higher concentrations of NT produces a cessation of measurable activity [36, 37]. However, it is not clear if this cessation of activity is related to toxic effects or the induction of 'depolarization block'. Electrophysiological investigations utilizing NT fragments are consistent with previous neurochemical and behavioral data supporting the C-terminal fragment as the neuroactive portion of the neuropeptide. Coapplication of a variety of other neurotransmitter receptor agonists and antagonists including GABA, glutamate, substance P, NMDA, and proglumide (CCK receptor antagonist) have not demonstrated any alterations in NT-induced excitations. Intracerebroventricular NT (50 μg) attenuated the inhibitory effects of systemically administered (i.v.) quinpirole, a specific D_2-agonist, on midbrain DA neurons [38]. NT has also been shown to increase the IC_{50} of BHT 920, a dopamine agonist. Directly applied NT, like parentally applied antipsychotic drugs attenuates DA-induced inhibitions in the VTA and SN [37]. It is noteworthy that NT fragments that do not bind DA also attenuate DA-induced inhibitions [39].

Pharmacological

Preclinical studies in rats that have utilized a variety of typical antipsychotic drugs (haloperidol, chlorpromazine, and pimozide) demonstrate significant increases in NT concentrations in specific brain regions, including the NA and caudate nucleus [40, 41] after acute and chronic administration. However, so called atypical antipsychotics such as clozapine and S(+)-N-propylnorapomorphine increase NT concentrations in the NA, whereas others, BMY 14802 and CI-943 do so in both the NA and the caudate nucleus (fig. 1) [41–44]. In contrast, phenothiazines without antipsychotic efficacy (promazine, promethazine) do not produce such an increase [40]. The selectivity of the increases in NT concentration associated with acute and chronic antipsychotic drug treatment is confirmed by the ineffectiveness of anxiolytics, anticonvulsants, and antihistamines in producing any significant changes in NT concentrations in these brain regions [45]. Haloperidol and BMY 14802 also decrease the NT concentrations in the prefrontal cortex (see above). The effects of clozapine on NT concentrations in the nucleus accumbens and its lack of effect in the caudate may be attributable to its lack of extrapyramidal side effects. Chronic (8 months) treatment with haloperidol increases NT in these areas [46], indicating that like the therapeutic effects of antipsychotic drugs, tolerance

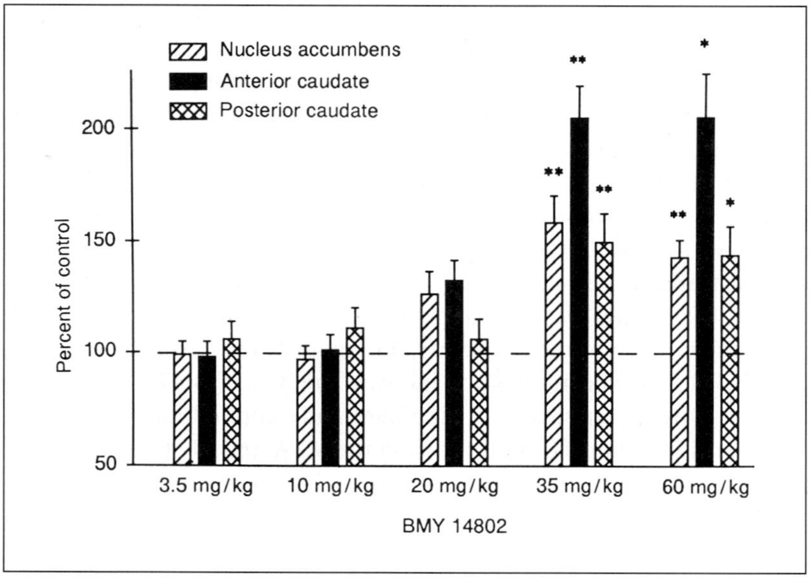

Fig. 1. Dose-response effects of BMY 14802 on regional NT concentrations. Animals were treated with BMY 14802 and sacrificed 18 h after injection. Values represent means ± SE expressed as percentage of control; n = 7–8 animals; ** $p < 0.01$; * $p < 0.05$ by ANOVA and Dunnett's test [42].

does not develop to this neurochemical action. The abrupt withdrawal of haloperidol results in a decrease in NT concentrations below baseline. The increase in NT concentrations in the NA and caudate associated with both haloperidol and BMY 14802 treatment is preceded by an increase in the expression of proNT mRNA as measured by in situ hybridizations (fig. 2) [47–49].

Interestingly, clozapine treatment increases NT mRNA expression only in the NA whereas haloperidol in contrast increases NT mRNA expression in both the NA and caudate nucleus confirming previous measurements. Similar results have recently been obtained with the atypical antipsychotic remoxipride [D. Dorsa, personal commun.]. In a series of studies utilizing a variety of nondopaminergic receptor antagonists, the effects of antipsychotic drugs on NT concentrations appear to be mediated, at least in part, by their effects on DA [7 for review]. Moreover, treatment with the selective D_1-receptor antagonist, SCH 23390, had no

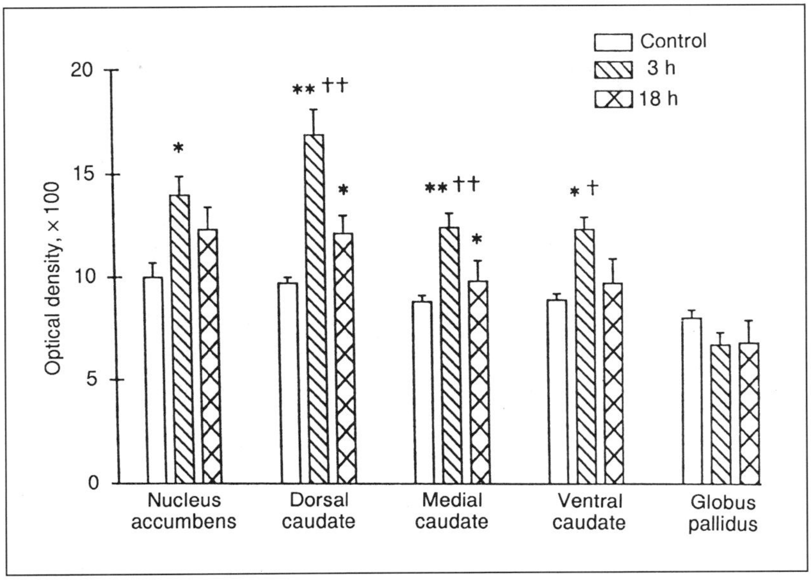

Fig. 2. Quantification of the effects of BMY 14802 on regional levels of proneurotensin mRNA. Rats (n = 5 per group) were sacrificed 3 and 18 h after treatment with BMY 14802 (35 mg/kg, i.p.). Autoradiograms were quantified with a RAS-1000 image analysis system. Data is expressed as average optical density × 100 per region sampled; ** $p < 0.01$; * $p < 0.05$ when compared to control; $^{+}p < 0.01$, $^{++}p < 0.05$ when compared to 18 h by ANOVA and the Student-Newman-Keuls multiple comparisons test [44].

effect on BMY 14802 and haloperidol-induced alterations in NT concentrations [50]. However, coadministration of SCH 23390 did attenuate the sulpiride-induced increase in NT in the caudate nucleus. These findings raise several questions whether these antipsychotic drug-induced alterations in NT neurons mediate, at least in part, the therapeutic effects of these drugs.

Clinical

Several clinical studies have repeatedly demonstrated a mean reduction in cerebrospinal fluid (CSF) concentrations of NT in medication-free patients with schizophrenia [51–53]. These reductions have not been observed in patients with major depression, premenstrual syndrome, and bulimia [54]. CSF NT concentrations increase following treatment with antipsychotic drugs, a result not unexpected considering the preclinical studies summarized above. The magnitude of the CSF NT reduction was positively correlated with decreased motor activity [52], increased hallucinations-delusions, thought disorder, and disorganized behavior [51]. Interestingly, CSF NT was lower in a subpopulation of patients that were lithium nonresponders, underscoring the potential selectivity for schizophrenic as compared to bipolar illness.

Finally, postmortem studies in patients with schizophrenia have demonstrated increased NT concentrations in Brodmann's area (BA) 32, but not in BA 12 or 24, nucleus accumbens, hypothalamus, or striatum [55].

Discussion

Clearly, antipsychotic drugs remain the cornerstone of the treatments available for schizophrenia. However, the utility of these agents is often hindered by the development of side effects and liability for tardive dyskinesia. It would be simplistic to believe that medications capable of treating a complex disorder were in themselves limited to only one neurotransmitter or neuropeptide system. There is currently the emergence of a more realistic appreciation of the intricacy of CNS neurotransmission including the D_1 and D_2 interactions as well as 5HT and DA interactions [4]. Another hypothesis involves the alteration of DA and neuropeptide ratios in the synaptic cleft [56].

In summary, scrutiny of the extant literature provides substantial evidence that NT may have a role in the mechanism of action of antipsychotic drugs. It has been hypothesized that NT may represent an endogenous antipsychotic substance [57]. Indeed, all typical and atypical antipsychotic drugs that have been studied thus far increase NT concentrations in certain brain regions. In addition, several putative antipsychotic substances including BMY 14802, S(+)-N-propylnorapomorphine, CI-943 also alter NT concentrations. The mechanism for this increase has not been conclusively delineated. Potential mechanisms for this increase include the increased biosynthesis of NT and/or the reduction of the activity of peptidases which degrade NT [58]. The receptors which mediate these alterations are unknown. One common receptor site for the increase in NT concentrations associated with antipsychotic drugs may potentially be the sigma receptor.

Yet the data surrounding the sigma receptor and potential subtypes have become increasingly complex. The effects of the putative antipsychotic CI-943 on NT concentrations provide additional evidence for the presence of a novel receptor mediating the NT response to antipsychotic drugs. Extensive receptor binding studies involving CI-943 have not identified any receptor to which CI-943 binds with high affinity. Unfortunately, no NT receptor agonists or antagonists have been developed. The cloning of the NT receptor [9] may provide such agents. The differential neurochemical effects in the mesolimbic and nigrostriatal DA systems produced with clozapine treatment suggest that lipophilic NT receptor agonists may be devoid of extrapyramidal side effects and tardive dyskinesia liability.

References

1 Creese I, Burt DR, Snyder SH: Dopamine receptor binding predicts clinical and pharmacological potencies of antischizophrenic drugs. Science 1976;192:481–483.

2 Seeman P, Chau-Wong M, Wong K: Antipsychotic drug doses and neuroleptic/dopamine receptors. Nature 1976;261:717–719.

3 Peroutka SJ, Snyder SH: Relationship of neuroleptic drug effects at brain dopamine, serotonin, alpha adrenergic, and histamine receptors to clinical potency. Am J Psychiatry 1980;137:1518–1522.

4 Deutch AY, Moghaddam B, Innis RB, Krystal JH, Aghajanian GK, Bunney SB, Charney DS: Mechanism of action of atypical antipsychotic drugs; implications for novel therapeutic strategies for schizophrenia. Schizophr Res 1991;4:121–156.

5 Leeman SE, Carraway RE: Neurotensin: Discovery, isolation, characterization, synthesis and possible physiological roles. Ann NY Acad Sci 1982;400:1–16.

6 Kitabgi P: Neurotensin modulates dopamine neurotransmission at several levels along brain dopaminergic pathways. Neurochem Int 1989;14:111–119.

7 Levant B, Nemeroff CB: The psychobiology of neurotensin. Curr Top Neuroendocrinol 1988;8:231–262.

8 Dobner PR, Barber DL, Villa-Komaroff LV, McKiernan C: Cloning and sequence analysis of cDNA for the canine neurotensin/neuromedin N precursor. Proc Natl Acad Sci USA 1987;84:3516–3519.

9 Tanaka K, Masu M, Nakaishi S: Structure and functional expression of the cloned rat neurotensin receptor. Neuron 1990;4:847–854.

10 Uhl GR, Snyder SH: Regional and subcellular distribution of brain neurotensin. Life Sci 1977;19:1827–1832.

11 Bissette G, Nemeroff CB, Decker MW, Kizer JS, Agid Y, Javoy-Agid F: Alterations in regional brain concentrations of neurotensin and bombesin in Parkinson's disease. Ann Neurol 1985;17:324–328.

12 Griffiths EC, McDermott JR, Smith AI: A comparative study of neurotensin inactivation by brain peptidases from different vertebrate species. Comp Biochem Physiol 1984;77C:363–366.

13 Iversen LL, Iversen SD, Bloom F, Douglas C, Brown M, Vale W: Calcium dependent release of somatostatin from rat brain in vitro. Nature 1978;273:161.

14 Jennes L, Stumpf WE, Kalivas PW: Neurotensin: Topographical distribution in rat brain by immunohistochemistry. J Comp Neurol 1982;210:211–224.

15 Nemeroff CB: The interaction of neurotensin with dopaminergic pathways in the central nervous system: basic neurobiology and implications for the pathogenesis and treatment of schizophrenia. Psychoneuroendocrinology 1986;11:15–37.

16 Bean AJ, During MJ, Deutch AY, Roth RH: Dopamine and neurotensin storage in colocalized and non colocalized neuronal populations. J Pharmacol Exp Ther 1989; 249:681–687.

17 Studler J-M, Kitabgi P, Tramu G, Herve D, Glowinski T, Tassin J-P: Extensive colocalization of neurotensin with dopamine in rat meso-cortico-frontal dopaminergic neurons. Neuropeptides 1988;11:95–100.

18 Palacios JM, Kuhar MJ: Neurotensin receptors are located on dopamine-containing neurons in rat midbrain. Nature 1981;294:587–589.

19 Quirion R, Everist HD, Pert A: Nigrostriatal dopamine terminals bear neurotensin receptors but the mesolimbic do not. Soc Neurosci Abstr 1982;8:582.

20 Goedert M, Mantyh PW, Hunt SP, Emson PC: Mosaic distribution of neurotensin-like immunoreactivity in the cat striatum. Brain Res 1983;274:176–179.

21 Herve D, Tassin JP, Studler JM, Dana C, Kitabgi P, Vicent JP, Glowinski J, Rostene W: Dopaminergic control of ^{125}I-labeled neurotensin binding site density in corticolimbic structures of the rat brain. Proc Natl Acad Sci USA 1986;83:6203–6207.

22 Reches A, Burke RE, Jiang D, Wagner HR, Fahn S: The effects of neurotensin on dopaminergic neurons in rat brain. Ann NY Acad Sci 1982;400:420–421.

23 Meyers RD, Lee TF: Neurotensin perfusion of rat hypothalamus: Dissociation of dopamine release from body temperature change. Neuroscience 1984;12:241–253.

24 Meyers RD, Lee TF: In vivo release of dopamine during perfusion of neurotensin in the substantia nigra of the unrestrained rat. Peptides 1984;4:955–961.

25 Nemeroff CB, Cain ST: Neurotensin-dopamine interactions in the central nervous system. Trends Pharmacol Sci 1984;6:201–205.

26 Aganti LF, Fuxe K, Benefenti F, Battistini N: Neurotensin in vitro markedly reduces the affinity in subcortical limbic [^{3}H]-N-propylnorapomorphine binding sites. Acta Physiol Scand 1983;119:459–461.

27 Von Euler G, Fuxe K: Neurotensin reduces the affinity of D_2 dopamine receptors in rat striatal membranes. Acta Physiol Scand 1987;131:625–626.

28 Nemeroff CB, Bissette G, Manberg PJ, Osbahr AJ III, Breese GR, Prange AJ Jr: Neurotensin-induced hypothermia: evidence for an interaction with dopaminergic systems and the hypothalamic-pituitary-thyroid axis. Brain Res 1980;195:69–84.

29 Bissette G, Nemeroff CB, Loosen PT, Breese GR, Burnett GB, Lipton MA, Prange AJ: Modification of pentobarbital-induced sedation by natural and synthetic peptides. Neuropharmacology 1978;17:229–237.

30 Ervin GN, Birekemo LS, Nemeroff CB, Prange AJ Jr: Neurotensin blocks certain amphetamine behaviors. Nature 1981;88:837–846.

31 Luttinger D, Wiggins R, King AR, Nemeroff CB, Prange Jr AJ: Inhibition of conditioned avoidance responding and rewarding electrical self-stimulation of the brain by centrally administered neurotensin. Soc Neurosci Abstr 1980;6:843.

32 Glimcher PW, Giovino AA, Hoebel BG: Neurotensin self-injection in the ventral tegmental area. Brain Res 1987;403:147–150.
33 Stowe ZN, Nemeroff CB: The electrophysiological actions of neurotensin. Life Sci 1991;49:987–1002.
34 Andrade R, Aghajanian GK: Neurotensin selectively activates dopaminergic neurons in the substantia nigra. Soc Neurosci Abstr 1981;7:184.1.
35 Pinnock RD: Neurotensin depolarizes substantia nigra dopamine neurons. Brain Res 1985;338:151–154.
36 Seutin V, Massotte L, Dresse A: Electrophysiological effects of neurotensin on dopaminergic neurons of the ventral tegmental area in the rat in vitro. Neuropharmacology 1989;9:949–954.
37 Shi W-X, Bunney BS: Neurotensin modulates autoreceptor mediated dopamine effects on midbrain dopamine cell activity. Brain Res 1991;543:315–321.
38 Shi W-X, Bunney BS: Neurotensin attenuates dopamine D2 agonist quinpirole induced inhibition of midbrain dopamine neurons. Neuropharmacology 1990;29: 1095–1097.
39 Shi W-X, Bunney BS: Effects of neurotensin on midbrain dopamine neurons: Are they mediated by formation of a neurotensin-dopamine complex. Synapse 1991;9: 157–164.
40 Govoni S, Hong JS, Yang H-YT, Costa E: Increase of neurotensin content elicited by neuroleptics in nucleus accumbens. J Pharmacol Exp Ther 1980;215:413–417.
41 Kilts CD, Anderson CM, Bissette G, Ely TD, Nemeroff CB: Differential effects of antipsychotic drugs on the neurotensin content of discrete brain nuclei. Biochem Pharmacol 1988;37:1547–1554.
42 Levant B, Nemeroff CB: Sigma receptor 'antagonist' BMY 14802 increases neurotensin concentrations in the rat nucleus accumbens and caudate. J Pharmacol Exp Ther 1990;254:330–335.
43 Nemeroff CB, Kilts CD, Levant B, Bissette G, Campbell A, Baldessarini RJ: Effects of the isomers of N-n-propylnorapomorphine and haloperidol on regional concentrations of neurotensin in rat brain. Neuropsychopharmacology 1991;4:27–34.
44 Levant B, Bissette G, Davis D, Heffner TG, Nemeroff CB: Effects of CI-943, a potential antipsychotic drug, and haloperidol on regional brain neurotensin concentrations. Synapse 1991;9:225–230.
45 Frey P, Fuxe K, Eneroth P, Aganti LF: Effects of acute and long-term treatment with neuroleptics on regional telencephalic neurotensin levels in the male rat. Neurochem Int 1986;8:429–434.
46 Radke JM, MacLennan AJ, Beinfeld MC, Bissette G, Nemeroff CB: Effects of short- and long-term haloperidol administration and withdrawal on regional brain cholecystokinin and neurotensin concentrations in the rat. Brain Res 1989;480:178–183.
47 Williams TG, Murtaugh MP, Beitz AJ: The effects of acute haloperidol treatment on brain proneurotensin mRNA: In situ hybridization analysis using a novel fluorescence detection procedure. Mol Brain Res 1990;7:347–358.
48 Merchant KM, Miller MA, Ashleigh EA, Dorsa DM: Haloperidol rapidly increases the number of neurotensin mRNA-expressing neurons in neostriatum of the rat brain. Brain Res 1991;540:311–314.

49 Levant B, Merchant KM, Dorsa DM, Nemeroff CB: BMY 14802, a potential antipsychotic drug increases expression of proneurotensin mRNA in the rat striatum. Mol Brain Res 1992;12:279–284.

50 Levant B, Nemeroff CB: Further studies on the modulation of regional brain neurotensin concentrations by antipsychotic drugs: Focus on halperidol and BMY 14802. J Pharmacol Exp Ther, in press.

51 Garver DL, Bissette G, Yao JK, Nemeroff CB: CSF neurotensin concentrations in psychosis: Relationship to symptoms and drug response. Am J Psychiatry 1991;148: 485–488.

52 Lindstrom LH, Widerlov E, Bissette G, Nemeroff CB: Reduced CSF neurotensin concentration in drug-free schizophrenic patients. Schizophr Res 1988;1:55–59.

53 Widerlov E, Lindstrom LH, Bresev G, Manberg PJ, Nemeroff CB, Breese GR, Kizer JS, Prange AJ Jr: Subnormal CSF levels of neurotensin in a subgroup of schizophrenic patients: Normalization after neuroleptic treatment. Am J Psychiatry 1982; 139:1122–1126.

54 Manberg PJ, Nemeroff CB, Bissette G, Prange AJ Jr, Gerner RH: Cerebrospinal fluid levels of neurotensin-like immunoreactivity in normal controls and in patients with affective disorder, anorexia nervosa and premenstrual syndrome. Soc Neurosci Abstr 1983;9:1054.

55 Nemeroff CB, Youngblood W, Manberg PJ, Prange AJ Jr, Kizer JS: Regional brain concentrations of neuropeptides in Huntington's chorea and schizophrenia. Science 1983;221:972–975.

56 Gariano RF, Groves PM: A mechanism for the involvement of colocalized neuropeptides in the actions of antipsychotic drugs. Biol Psychiatry 1989;26:303–314.

57 Nemeroff CB: Neurotensin: Perchance an endogenous neuroleptic? Biol Psychiatry 1980;15:283–302.

58 Morril M, Louis R, Davis TP: Haloperidol, chlorpromazine, and apomorphine alter central regional neuropeptidase activity. Eur J Pharmacol 1990;183:2318–2319.

Charles B. Nemeroff, MD, PhD, Department of Psychiatry, Emory University School of Medicine, 1701 Uppergate Drive, Atlanta, GA 30322 (USA)

Brunello N, Mendlewicz J, Racagni G (eds): New Generation of Antipsychotic Drugs: Novel Mechanisms of Action. Int Acad Biomed Drug Res. Basel, Karger, 1993, vol 4, pp 27–38

What Are Atypical Neuroleptics and How Do They Work?

David M. Jackson, Nina Mohell, Anneli Bengtsson, Åsa Malmberg[1]

Departments of Behavioral and Neuropharmacology, Preclinical CNS Research and Development, Astra Arcus AB, Södertälje, Sweden

Introduction

Since the introduction of the dopamine (DA) receptor antagonists (neuroleptics) as an effective therapy for various psychoses including schizophrenia, it has become evident that some of them (e.g., clozapine, sulpiride and remoxipride) produce fewer extrapyramidal side effects (and possibly less tardive dyskinesia) than others (e.g., haloperidol). All, however, are efficacious provided that an adequate dose and treatment schedule is used. In addition, some (e.g., clozapine) may be more effective than others in the treatment of resistant patients. The neuroleptics that produce fewer extrapyramidal side effects (and which may be more effective in 'resistant' patients) have been termed atypical. This clinical definition has been supplemented by a preclinical definition supported by preclinical models. In this paper, an atypical neuroleptic (actual or potential) is defined as one that has, on the one hand, a low propensity to cause catalepsy in rodents but, on the other hand, a potent ability to block the hyperlocomotion produced by drugs such as *d*-amphetamine. In contrast, a typical neuroleptic is one that is about equipotent in both of the behavioral models. The degree of catalepsy may be predictive of the likely incidence of extrapyramidal problems in man and is caused by DA D_2-receptor blockade within the basal ganglia. Blockade of *d*-amphetamine-induced hyperlocomotion is considered to be predictive of neuroleptic efficacy and results from the blockade of DA D_2-receptors within the mesolimbocorti-

[1] Annica Rask is thanked for excellent assistance.

Table 1. Ability of some typical and atypical neuroleptics to produce catalepsy and to block the locomotor stimulant effect of *d*-amphetamine in rats

	Catalepsy $ED_{0.5}$	Amphetamine blockade ED_{50}	Ratio of ED values cat/amphet	Atypical in clinic
Clozapine	>120	2.4	>50.6	Yes
Chlorpromazine	>120	0.5	>42	?
SCH23390	1.8	0.04	40.5	?
Remoxipride (–)	136	6.7	20.4	Yes
Risperidone	13.1	0.8	16.8	Yes (?)
Raclopride (–)	4.3	0.8	5.5	Yes (??)
Sulpiride (–)	>480	136	>3	Yes
Spiperone	0.1	0.06	2.3	No
Haloperidol	0.3	0.2	1.8	No

Catalepsy; Rats were injected s.c. with various doses of a neuroleptic and the catalepsy (time spent immobile on an inclined plane) measured several times between 0.5 and 24 h after injection. The $ED_{0.5}$ for catalepsy is the average dose required to produce catalepsy (at the time of peak effect) for 0.5 min. $ED_{0.5}$ values in µmol/kg after s.c. administration.
D-Amphetamine test; Rats were injected s.c. with various doses of neuroleptics and 1 h later with *d*-amphetamine (5 µmol/kg) and the total activity measured and accumulated for 1 h. The dose of neuroleptic (µmol/kg) required to reduce the response to 50% of *d*-amphetamine alone was calculated as theED_{50}.
Ratios; The ratio of ED values, cat/amphet is the ratio of the catalepsy $ED_{0.5}$ value to the amphetamine blockade ED_{50}.

cal system, especially the nucleus accumbens. The preclinical definition should *not* be confused with the clinical definition. The evidence that neuroleptics that are atypical in animals are also atypical in the clinic (while maintaining efficacy) is not strong. Some neuroleptics are illustrated in table 1. Drugs like clozapine, remoxipride and perhaps risperidone seem to be atypical both in the clinic and in preclinical testing (table 1). The atypicality is evidenced by a high ratio of $ED_{0.5}$ (catalepsy) to ED_{50} (*d*-amphetamine) – clozapine and remoxipride, for example, exhibit ratios of about 50 and 20, respectively. Thus, they have a low tendency to cause catalepsy relative to their ability to block amphetamine. In contrast, haloperidol and spiperone, which are not atypical in the clinic and do not exhibit atypicality in the preclinical tests (table 1), display ratios of about 2.

Despite the differing side effect potential of the various drugs, DA D_2-receptor blocking activity is an absolute requirement of neuroleptic efficacy. There is a positive correlation between the affinity of a neuroleptic for the D_2-receptor and clinical efficacy [1], which does not occur for any other receptor affinity. However, it is true that activity at receptors other than DA D_2 can modulate either favorably or unfavorably a neuroleptic's side effect profile (and probably efficacy). For example, chlorpromazine's low incidence of extrapyramidal side effects may be partly dependent upon its antimuscarinic properties, which also (unfortunately) cause some of the peripheral side effects associated with this drug.

The aim of the present article is not to discuss the entire pharmacology of typical and atypical neuroleptics. The field is huge with hypotheses spanning not only the DA area but also, for example, σ-binding sites, excitatory amino acid neurotransmitters and $5HT_3$-antagonists. Most of these areas will not be covered in the present report and the reader is referred to reviews in these areas [2–10]. This report will discuss only a few approaches that have been and are being used in an attempt to develop new and atypical neuroleptics.

Mechanisms Underlying Atypical Neuroleptics

Receptor Affinity Profile

At first glance, a particular receptor profile cannot explain atypicality. Table 2 lists receptor affinities for a few typical and atypical neuroleptics. Consider clozapine, remoxipride and risperidone. Clozapine exhibits its highest affinity to $5HT_2$ receptors (K_i, 13.3 n*M*), and less for DA D_2 (83 n*M*) and DA D_1 (174 n*M*) receptors (table 2). It also has rather high affinity for α_1-adrenergic and for M_1- through M_5-muscarinic receptors [11] – this muscarinic affinity could be important (see below). In pharmacological terms, clozapine is a dirty drug. In contrast, remoxipride has a cleaner receptor profile with similar, but low, affinities for DA D_2 rat striatal receptors (240 n*M*) and σ-binding sites (268 n*M*) and very little affinity at any other receptors. In contrast to clozapine and remoxipride, risperidone has high affinities for $5HT_2$ (0.2 n*M*), α_1 (0.81 n*M*) and DA D_2 (3.1 n*M*) receptors, with a much lower affinity for DA D_1 (369 n*M*) receptors. Yet clozapine, remoxipride and risperidone have all been reported to be atypical neuroleptics, suggesting that $5HT_2$-, σ- and α_1-adrenergic affinities per se are not *absolutely* indispensable for neuroleptic activity. From

Table 2. Binding affinities of some typical and atypical neuroleptics for various receptors (data represent the K_i values in n*M* unless otherwise stated)

	D_2[1]	D_1[2]	$5HT_2$[3]	σ[4]	HT_{1a}[5]	α_1[6]
Chlorpromazine	4.1		24.9	1,080		14
Clozapine	83	174	13.3	10,500	>3,000	15
SCH23390	266	2.4				
Remoxipiride (–)	240	250,000	>22,000[7]	268	>81,000[8]	42,000
Risperidone[9]	1.2	269		852		20.81[11]
	3.10[10]	534	0.2		253	
Raclopride (–)	1.3	>54,700	>11,000	>14,300		32,000
Sulpiride (–)	1.9	>17,500				54,000
Spiperone	0.1	714		1,680		28
Haloperidol	0.3	35.1	34	2.7		35
Sertindole[12]	4.1	28	0.49	88	2,600	3.4
Amperozide[13]	540	2,900	16.5		805	172

[1] (^{3}H)-raclopride to rat striatal membranes, except for some risperidone data, sertindole and amperozide.
[2] (^{3}H)-SCH23390 to rat striatal membranes, except for risperidone, sertindole and amperozide.
[3] (^{3}H)-ketanserin to rat cortical membranes, except for risperidone, sertindole and amperozide.
[4] (^{3}H)-DTG to guinea pig whole brain membranes, except for risperidone, sertindole and amperozide.
[5] (^{3}H)-8-OHDPAT to rat hippocampal (risperidone) or cortical (all other drugs) membranes.
[6] All values are IC_{50} values from (^{3}H)-WB4101 binding in rat cortex except for sertindole and amperozide.
[7] (^{3}H)-5HT to rat striatal membranes.
[8] (^{3}H)-spiperone to rat striatal membranes.
[9] See Leysen et al. [46].
[10] (^{3}H)-8-haloperidol to rat striatal membranes.
[11] (^{3}H)-WB4101 to rat cortical membranes.
[12] IC_{50} values from Arnt et al. [45].
[13] Rat brain K_i values are from Svartengren and Simonsson [47]. $5HT1_{1a}$, (^{3}H)-8-OHDPAT in hippocampus; $5HT_2$, (^{3}H)-ketanserin to cortex; α_1, (^{3}H)-prazosin to cortex: D_1, (^{3}H)-SCH23390 to striatum; D_2, (^{3}H)-raclopride to striatum

these few examples [others are discussed in 12] it is clear that no particular receptor profile will with predictable results confer atypicality on a neuroleptic. However, a word of caution: it is always possible that atypicality is dependent in some cases on *the ratio* of the various affinities for the different receptors, and that a compound such as clozapine or risperidone is atypical because it has exactly the right balance between these affinities. One final point. The high affinity of clozapine for various muscarinic receptor subtypes is possibly very important, since muscarinic antagonists have been known for centuries to produce psychiatric disorders, and acetylcholine as a neurotransmitter is involved in cognitive functions. It is possible that the atypicality seen with clozapine is in some measure dependent upon this muscarinic affinity and it shall be interesting to see the effect in man of selective M_1- through M_5-agonists and antagonists.

σ-Affinity

σ-Binding sites were first defined as a way of explaining the psychotomimetic effects of certain opiates and related drugs [13]. Certain of these effects resembled some of the symptoms seen in psychoses such as schizophrenia. This finding led to the concept that at least some σ-binding site ligands may have desirable central nervous system effects intrinsically or at least influence in a positive way (better efficacy, fewer side effects) DA D_2-receptor-mediated effects. As a result, many substances with σ-site binding activity have been synthesized. Such compounds have σ-binding site affinity either as a dominant feature of the molecule or as one of several receptor affinities. In addition, a large variety of other substances, some of which are or may be neuroleptics (remoxipride [14], umespirone [15], and haloperidol together with some of its metabolites [16] or other types of compounds such as 3PPP (3-(3-hydroxyphenyl)-N-n-propylpiperidine) display significant σ-site binding affinities. While it cannot yet be ruled out that a σ-binding site may ultimately be shown to play a role in therapeutic actions of neuroleptics, it remains an open question. In part, the resolution of the uncertainties in the field will depend upon the synthesis of highly selective agonists and antagonists. However, to date, even substances like rimcazole (BW234U) and BMY14802, which have been suggested to be relatively selective σ-ligands, display significant affinities for other receptors. Thus, BMY14802 has a high affinity for $5HT_{1a}$ receptors. This is important because BMY14802 fails to induce catalepsy and this may be due to its $5HT_{1a}$ affinity. [For up-to-date general reviews dealing with the σ-field, see 17, 18]. In summary, there are problems with the hypothesis and these include: (1) the inability today to define substances as agonists or antagonists and the ignorance at present of an endogenous ligand (if there is one); (2) an ignorance about the number of σ-binding subtypes in the CNS, with ligands such as DTG, 3PPP and SKF10047 labelling multiple sites; (3) the lack of evidence that σ-binding size represents a functional receptor [18 for discussion] and not merely an acceptor site or a protein or enzyme binding site. Thus, it seems likely that some σ-sites may be associated with cytochrome P450 [19, 20] and with the NMDA ion channel [21]; (4) the lack of correlation between σ-binding site affinity and either efficacy or incidence of side effects seen with neuroleptics (both haloperidol and remoxipride have σ-binding site and D_2-receptor affinity (table 2); (5) many compounds that have no obvious neuroleptic activity have high σ-binding activity – for example a number of rather selective monoaminoxidase type A inhibitors have high σ-affinity and presumably no neuroleptic activity [22]. Resolution of the importance of σ-binding sites in conferring atypicality or otherwise must await further basic research in the σ-area. On the other hand, evidence is now emerging that some of the motor effects of some neuroleptics may be partly mediated via σ-binding sites [see 23 for review and 24].

DA Receptor Subtypes

It was rather early suggested that neuroleptics may function on subtypes of DA receptors and it has been suggested that the substituted benzamides such as sulpiride may even selectively target subtypes of D_2 receptor [25, 26]. The evidence has, however, been indirect. Nevertheless, just when most researchers in the DA field had thought that the DA receptor story had stabilized to two receptor groups, D_1 and D_2, each with high and low affinity states, molecular biology has provided us with splice variants of the D_2 receptors (D_{2_A}, long and D_{2_B}, short), together with two more DA receptors, D_3 and D_4, which are both closely related to the D_2 receptor in their sequences [27–29]. DA D_3 and D_4 receptors are potentially interesting because their distribution is different from the distribution of D_2 receptors with some location in limbic areas. Their description has provoked discussion as to whether they may be suitable targets for neuroleptics. Indeed, clozapine has been described as selective for the D_4 receptors [29]. We have examined the binding profile of neuroleptics on cloned DA D_3 receptors, as well as on D_{2_A} and D_{2_B} receptors (table 3). At the moment, there is not enough evidence to clearly demonstrate if a particular affinity for a D_2 or D_3 receptor subtype can explain atypicality. It is interesting, however, that remoxipride has a considerably higher affinity for human D_2 than for D_3 (table 3) receptors, in contrast to raclopride, which has about equal affinity for human D_2 and D_3 receptors (table 3). Although the emphasis is on obtaining selective D_3 and D_4 drugs, it is equally as interesting to obtain selective D_2 drugs.

Table 3. Binding affinities of some neuroleptics and SCH23390 at cloned DA D_2 and D_3 receptors (mean ± SEM)

Neuroleptic	Affinity, (k_i, nM)		
	D_3 receptors	D_2 receptors	
		long (A)	short (B)
Clozapine	105 ± 19	65.7 ± 22.6	46.2 ± 6.7
Remoxipride	945 ± 63	159.4 ± 0.2	48.6 ± 2.1
Raclopride	2.1 ± 0.3	2.7 ± 0.2	1.1 ± 0.3
(–)-Sulpiride	10.3 ± 0.9	8.6 ± 0.01	2.7 ± 0.5
Haloperidol	2.6 ± 1.0	0.77 ± 0.11	0.5 ± 0.1
SCH23390	338 ± 85	267 ± 37	324 ± 5

Human DA D_{2A} and D_{2B} receptors expressed in mouse fibroblasts (Ltk^- cells). Human DA D_3 receptors expressed in Chinese hamster ovary (CHO) cells. (^3H)-raclopride used as the ligand in all studies. K_i values were calculated using the LIGAND program (n = 2–4).

In the same way that molecular biology is affecting D_2 receptor research, various D_1 receptor subtypes have been described – their functional significance is, however, unclear. Although affinity for D_1 receptors is not a prerequisite for neuroleptic activity, since many effective neuroleptics have little or no D_1 affinity (table 2), their pharmacology and possible clinical significance have stimulated a considerable amount of interest. On the one hand, clozapine has considerable affinity for D_1 receptors (table 2) and has even been suggested to bind to a subtype of D_1 receptor [30]. On the other hand, remoxipride has no significant DA D_1 receptor affinity (table 2). Although SCH23390 behaves in animal models as an atypical neuroleptic (table 1), this reflects the fact that D_1 and D_2 antagonists influence catalepsy via different neuronal pathways. However, regardless of hypotheses about the role of D_1 receptors in schizophrenia, it is not known yet if a *selective* DA D_1 antagonist will have any efficacy in schizophrenia. In one preliminary report SCH23390 was reported to be ineffective in schizophrenic patients, but this was an open short-term trial with few patients and with a SCH23390 dose that may have been inadequate [13]. In Gessa et al. [31], and in a study in Cebus monkeys [32], evidence was presented that D_1-antagonists may in fact be useful in neuroleptic-induced movement disorders. Furthermore, it was shown in another study [33] that the local implantation of SCH23390 into the forebrain of monkeys had detrimental effects on cognition; certainly an undesirable effect in schizophrenia. A test of the DA D_1 concept must await the advent of a suitable DA D_1 antagonist and solid clinical studies.

DA Autoreceptor Agonists, or Partial Agonists

Partial DA D_2 agonists such as 3PPP [34, 35], EMD23448 [36], HW165 [37], SDZ208-911 and SDZ208-912 [38] and a series of N-cyclohexylbenzamides [39], amongst others, have been advanced as potential neuroleptics with the possibility of having a diminished side effect profile [34]. DA agonists (especially partial agonists) by stimulating DA autoreceptors reduce DA overflow into the synapse. Thus, the end result at the postsynaptic DA receptors is similar to that seen after neuroleptics – a net reduction in DA neurotransmission. The hope is, however, that DA autoreceptor agonists will cause a more controlled reduction in DA neurotransmission than that seen with traditional DA D_2 antagonists. Partial agonists can function as antagonists or agonists depending on the amount of intrisic activity, a property largely dependent upon the number of spare receptors. The concept that such a drug will be an effective antipsychotic with fewer side effects remains to be properly tested in the clinic, although some early uncontrolled studies have indicated a certain efficacy in schizophrenia. If a partial DA D_2 agonist is coupled in the same molecule with $5HT_{1a}$ agonist activity, such as occurs in roxindole (EMD49980) [40], then a prediction of the end clinical efficacy and side effect profile becomes all the more difficult. The key problems, perhaps, with the DA autoreceptor agonist hypothesis as it applies to selective DA D_2 ligands are; (1) will a reduction of 60–70% in DA overflow (the maximum achievable) be adequate for an antipsychotic effect?; (2) will the reduction of DA overflow onto both D_1 and D_2 receptors be a desirable property, given that there is no evidence that blockade of D_1 receptors plays a positive role in neuroleptic action and indeed may be deleterious (see discussion above)?

Electrophysiological Approaches

Neuroleptics such as haloperidol increase the firing rates of A8, A9 and A10 neurons. Clozapine, however, does not have a marked effect on A9 neurons. This observation, together with other data, indicates that the electophysiological properties of A8-A10 DA neurons differ from those of the A9 neurons. This brain-region-specific electrophysiological difference has been used as a method for the development of drugs selective for the A8-A10 DA system. How has this been achieved? Typical neuroleptics given chronically reduce the activity of both A8-A10 and A9 neurons. However, in early papers it was observed that chronic clozapine seemed to have a selective effect on the A8-A10 DA neuronal systems with fewer effects on the A9 system [41, 42]. A similar property has been shown with rimcazole (see above) [43]. Furthermore, 28 days of treatment of rats with BMY14802, a compound related to rimcazole, has been shown to produce a selective inactivation of A8-A10 but not of A9 neurons [44]. Finally, the putative neuroleptic sertindole seems to have been developed using this concept of electrophysiological selectivity [45]. Thus, sertindole is not selected in standard neuroleptic screening models, failing for example to antagonize methamphetamine or amphetamine-induced effects. In terms of receptor affinities, it exhibits high affinity for $5HT_2$ receptors and less for DA D_2 and D_1 receptors (table 2). However, in electrophysiological studies, A8-A10 neurons were inactivated by very low chronic doses whereas much higher chronic doses were required to inhibit the A9 neurons. Thus, sertindole in some ways resembles clozapine. The reason underlying this selective effect on A8 and A10 neurons is unknown. Some of the drugs studied using this electrophyisological approach are, receptor-wise dirty drugs, and it is not known how a certain receptor profile will influence the electrophysiological actions. While clozapine is certainly clinically active, the validity of this most interesting approach to atypical neuroleptic development must await results from clinical tests in patients with newer drugs such as sertindole.

Conclusions

Only a few approaches used in the development of atypical neuroleptics have been discussed here. In our opinion, it remains unclear if any one special property renders a neuroleptic atypical. Neuroleptics from a broad range of chemical classes which display no consistent receptor binding profile or other pharmacological properties have been shown to be atypical. By studying clozapine, clues have been provided that some atypical drugs may preferentially affect mesolimbic instead of basal ganglia DA systems. However, a planned replacement for clozapine, which has considerable toxic properties, remains elusive.

References

1 Seeman P, Lee T, Chau-Wong M, Wong K: Antipsychotic drug doses and neuroleptic/dopamine receptors. Nature 1976;261:717–719.

2 Creese I: Classical and atypical antipsychotic drugs: new insights. Trends Neurosci 1983;6:479–481.

3 Andersen PH, Nielsen EB: Novel approaches to development of antipsychotics. Drug News Perspect 1991;4:150-157.

4 Baldessarini RJ, Frankenburg FR: Clozapine. A novel antipsychotic agent. N Engl J Med 1991;324:746–754.

5 Vinick FJ, Heym JH: Antipsychotic agents; in Bailley DM (ed): Annual Reports in Medicinal Chemistry. New York Academic Press, 1987, vol 22, pp 1–10.

6 New JS, Takaki KS: Antipsychotic agents, in Bailey DM (ed): Annual Reports in Medicinal Chemistry. New York, Academic Press, 1988, vol 23, pp 1–10.

7 Högberg T: Novel substituted salicylamides and benzymides as selective dopamine D2-receptor antagonists. Drugs Future 1991;16:333–357.

8 Deutsch SI, Mastropaolo J, Schwartz BL, Rosse RB, Morihisa JM: A 'glutamatergic hypothesis' of schizophrenia. Rationale for pharmacotherapy therapy with glycine. Clin Neuropharmacol 1989;12:1–13.

9 Wachtel H, Turski L: Glutamate: A new target in schizophrenia? Trends Pharmacol Sci 1990;11:219–220.

10 Abbott A: 5-HT3 antagonists and ligands for dopamine D_1- and autoreceptors offer new leads for antipsychotic drugs. Trends Pharmacol Sci 1990;11:49–51.

11 Boplden C, Cusack B, Richelson E: Clozapine is a potent and selective muscarinic antagonist at the five cloned human muscarinic acetylcholine receptors expressed in CHO-K1 cells. Eur J Pharmacol 1991;192:205–206.

12 Seeman P: Atypical neuroleptics: Role of multiple receptors, endogenous dopamine, and receptor linkage. Acta Psychiatr Scand 1990;358:14–20.

13 Martin WR, Eades CG, Thompson JA, Huppler RE, Gilbert PE: The effects of morphine- and nalorphine-like drugs in the nondependent and morphine-dependent chronic spinal dog. J Pharmacol Exp Ther 1976;197:517–532.

14 Köhler C, Hall H, Magnusson O, Lewander T, Gustafsson K: Biochemical pharmacology of the atypical neuroleptic remoxipride. Acta Psychiatr Scand 1990;358:27–36.

15 Itzhak Y, Ruhland M, Krähling H: Binding of umespirone to the σ-receptor: Evidence for multiple affinity states. Neuropharmacology 1990;29:181–184.

16 Bowen WD, Moses EL, Tolentino PJ, Walker JM: Metabolites of haloperidol display preferential activity at sigma receptors compared to dopamine D-2 receptors. Eur J Pharmacol 1990;177:111–118.

17 Chavkin C: The sigma enigma; biochemical and functional corrrelates emerge for the haloperidol-sensitive sigma binding site. Trends Pharmacol Sci 1990;11:213–215.

18 Kaiser C, Pontecorvo MJ, Mewshaw RE; Sigma receptor ligands: function and activity. Neurotransmission 1991;7:1–5.

19 Ross SB: Is the sigma opiate receptor a proadifen-sensitive subform of cytochrome P-450. Pharmacol Toxicol 1990;67:93–94.

20 Ross SB: Heterogeneous binding of sigma radioligands in the rat brain and liver:

Possible relationship to subforms of cytochrome P-450. Pharmacol Toxicol 1990;68: 293–301.

21 Itzhak Y, Hiller JM, Simon EJ: Characterization of specific binding sites for (^{3}H):(d)-N-allnormetazocine in rat brain membranes. The search for selective dopaminergic autoreceptor agonists. Mol Pharmacol 1985;27:45–52.

22 Itzhak Y, Stein I: Sigma binding sites in the brain; an emerging concept for multiple sites and their relevance for psychiatric disorders. Life Sci 1990;47:1073–1081.

23 Walker JM, Bowen WD, Walker FO, Matsumoto RR, De Costa B, Rice KC: Sigma receptors: Biology and function. Pharmacol Rev 1990;42:355–402.

24 Walker JM, Matsumoto RR, Bowen WD, Gans DL, Jones KD, Walker FO: Evidence for a role of haloperidol-sensitive σ-'opiate' receptors in the motor effects of antipsychotic drugs. Neurology 1988;38:961–965.

25 Bischoff S, Bittinger H, Delini-Stula A, Ortmann R: Septo-hippocampal system: Target for substituted benzamides. Regional blockade by neuroleptic drugs of in vivo ^{3}H-spiperone binding in the rat brain. Relation to blockade of apomorphine induced hyperactivity and stereotypies. Eur J Pharmacol 1982;79:225–232.

26 Köhler C, Haglund L, Ögren S-O, Ängeby T: Regional blockade by neuroleptic drugs of in vivo ^{3}H-spiperone binding in that rat brain. Relation to blockade of apomorphine induced hyperactivity and stereotypies. J Neural Transm 1981;52:163–173.

27 Strange PG: Interesting times for the dopamine receptors. Trends Neurosci 1991;14: 43–45.

28 Sokoloff P, Giros B, Martres M-P, Bouthenet M-L, Schwartz J-C: Molecular cloning and characterization of a novel dopamine receptor D_3 as a target for neuroleptics. Nature 1990;347:146–151.

29 van Tol HHM, Bunzow JR, Guan H-C, Sunahara RK, Seeman P, Niznik HB, Civelli O: Cloning of the gene for a human dopamine D4 receptor with high affinity for the antipsychotic clozapine. Nature 1991;350:610–614.

30 Andersen PH, Braestrup C: Evidence for different states of the dopamine D1 receptor: Clozapine and fluperlapine may preferentially label an adenylate cyclase-coupled state of the D_1 receptor. J Neurochem 1986;47:1822–1831.

31 Gessa GL, Canu A, Del Zompo M, Burrai C, Serra G: Lack of acute antipsychotic effect of SCH23390, a selective dopamine D1 receptor antagonist. Lancet 1990;337: 854–855.

32 McHugh D, Coffin V: The reversal of extrapyramidal side effects with SCH39166, a dopamine D1 receptor antagonist. Eur J Pharmacol 1991;202:133–134.

33 Sawaguchi T, Goldman-Rakic PS: D1 dopamine receptors in prefrontal cortex: Involvement in working memory. Science 1991;251:947–950.

34 Clark D, Hjorth S, Carlsson A: Dopamine-receptor agonists: Mechanisms underlying autoreceptor selectivity. I Review of the evidence. The search for selective dopaminergic autoreceptor agonists. J Neural Transm 1985;62:1–52.

35 Clark D, Hjorth S, Carlsson A: Dopamine-receptor agonists: mechanisms underlying autoreceptor selectivity. II Theoretical considerations. J Neural Transm 1985;62: 171–207.

36 Seyfried CA, Fuxe K, Wolf HP, Agnati AF: Neurochemical and functional studies with EMD 23,488, a novel dopamine agonist. Acta Pharm Suec 1983(suppl 2):243–248.

37 Hjorth S, Svensson K, Carlsson A, Wikström H, Andersson B: Central dopaminergic properties of HW-165 and its enantiomers; *trans*-octahydrobenzo(f)quinoline congeners of 3-PPP. Naunyn Schmiedebergs Arch Pharmacol 1986;333:205–218.

38 Coward DM, Dixon AK, Urwyler S, White TG, Enz A, Karobath M, Shearman G: Partial dopamine-agonistic and atypical neuroleptic properties of the aminoergolines SDZ 208-911 and SDZ 208-912. J Pharmacol Exp Ther 1990;252:279–285.

39 Wise LD, Jaen JC, Caprathe BW, Smith SJ, Pugsley TA, Heffner TG: N-cyclohexylbenzamide: A novel class of dopamine autoreceptor agonists with potential antipsychotic activity. Soc Neurosci Abstr 1991;1:689.

40 Seyfried CA, Haase AF, Böttcher H: Presynaptic D2-receptor selectivity of roxindole (EMD 49 980): Possible indirect effects on postsynaptic striatal cholinergic neurons; in Tamminga CA, Shulz SC (eds): Schizophrenia – Advances in Neuropsychiatry. New York, Raven Press, 1991, vol 1, pp 317–322.

41 Chiodo LA, Bunney BS: Typical and atypical neuroleptics: differential effects of chronic administration on the activity of A9 and A10 midbrain dopamine neurons. J Neurosci 1983;3:1607–1619.

42 White FJ, Wang RY: Differential effects of classical and atypical antipsychotic drugs on A9 and A10 dopamine neurons. Science 1983;221:1054–1056.

43 Piontek JA, Wang RY: Acute and chronic effects of Rimcazole (BW 234U), a potential antipsychotic drug, on A9 and A10 dopamine neurons in the rat. Life Sci 1986; 39:651–658.

44 Wachtel SR, White FJ: Electrophsiological effects of BMY 14802, a new potential antipsychotic drug, on midbrain dopamine neurons in the rat. Acute and chronic studies. J Pharmacol Exp Ther 1988;244:410–416.

45 Arnt J, Skarsfeldt T, Meier E, Hyttel J, Perregaard J, Sanchez C: Sertindole – a new atypical neuroleptic. Soc Neurosci Abstr 1990;1:248.

46 Leysen JE, Gommeren W, Eens A, de Chaffoy de Courcelles D, Stoof JC, Janssen PAJ: Biochemical profile of risperidone, a new antipsychotic. J Pharmacol Exp Ther 1988;247:661–670.

47 Svartengren J, Simonsson P: Receptor binding properties of amperozide. Pharmacol Toxicol 1990;66(suppl 1):8–11.

David M. Jackson, Department of Behavioral and Neuropharmacology,
Preclinical CNS Research and Development, Astra Arcus AB,
S–151 85 Södertälje (Sweden)

Brunello N, Mendlewicz J, Racagni G (eds): New Generation of Antipsychotic Drugs: Novel Mechanisms of Action. Int Acad Biomed Drug Res. Basel, Karger, 1993, vol 4, pp 39–61

Origins of Psychosis and the Evolution of Human Language and Communication

Timothy J. Crow

Division of Psychiatry, Clinical Research Centre, Watford Road, Harrow, UK

It is widely assumed, perhaps particularly by those who are concerned with the prospects of introducing new treatments, that the underlying defect in schizophrenia could be a simple disturbance of neurohumoural function. In favour of this view is the fact that for 35 years we have had treatments that are relatively effective, and for the past 20 years or so a theory, the dopamine hypothesis, that accounts for this. I will argue that it is surprising that we have these treatments, given that our understanding of the disease process is so limited. One might say we hardly deserve to be so lucky. And we are unlikely to have the greater fortune of discovering radically new approaches unless we increase substantially our fundamental knowledge.

Consider the dopamine hypothesis around which so much research is orientated. The hypothesis can be formulated simply as that in affected individuals there is a deviation in transmitter enzyme or receptor function, leading to an excess of dopaminergic activity, that accounts for their symptoms. It is generally supposed that such a deviation would be genetic, i.e. associated with a mutation in a gene that codes for receptor or enzyme. But is it likely that the defect could be so simple? As far as we know the mechanisms for neurohumoural transmission are closely similar across at least the vertebrate phylum. The same mechanisms have been acted upon by evolutionary selective pressures for many millions of years. Why should such genes now be subject to important intraspecific variation in homo sapiens? If mutations in the gene for a neurotransmitter enzyme or receptor resulting in psychotic symptoms did arise one would expect, given the fertility disadvantage associated with such symptoms (see below), that the mutation would be rapidly eliminated. The paradox is that the disease persists at a high population frequency in spite of a fertility disadvantage. If we confine ourselves to psychoses that are predominantly schizophrenic the lifetime prevalence approaches 1%. If as I shall argue schizophrenic and affective psychoses should be considered together the prevalence is probably in excess of 2%. And yet these conditions (particularly schizophrenia) are associated with a substantial biological disadvantage.

Are There Environmental Factors in Causation?

Illness of high prevalence can be due to environmental causes. Infectious diseases are an example. But schizophrenia has epidemiological features that are shared by no infectious disease. The conclusions of the recent World Health Organisation ten country study are salutory – 'schizophrenic illnesses are ubiquitous, appear with similar incidence in different cultures and have clinical features that are more remarkable by their similarity across cultures than their difference' [1]. A number of environmental factors (viruses, toxins, allergens) have at one time or another been suggested as causes of schizophrenia, but few if any specific proposals remain as serious candidates. What environmental factor can one suggest that would not vary substantially across the globe? What factor could remain constant across wide variations in climate, industrial and social structure so as to affect a relatively constant proportion of individuals, and those in a particular phase of life?

Two further findings are relevant to the gene versus environment controversy. The first is that when the illness occurs in two siblings, it occurs at the same age, and not at the same time [2]. The second is that being adopted away in early childhood from a family in which there is already a member who suffers from schizophrenia, does not reduce risk of the disease [3]. Both findings suggest that environmental factors, at least those that occur in post-natal life, have little influence on the onset of schizophrenia. The possibility remains that an exogenous influence that impinges on the individual in fetal life or at the time of birth in some way predisposes that individual to schizophrenic illness 20 or more years later. Two such influences have been suggested – influenza in the mother in the second trimester (the 4th–6th months of pregnancy) and complications (e.g. hypoxia or head injury) occurring at the time of birth. Both hypotheses have recently been examined in a sample of 17,000 individuals whose births in a single week in 1958 and subsequent developments have been carefully documented (in the British Perinatal Mortality Survey and the National Child Development Study). We identified subjects who were later admitted to psychiatric units. Those who were diagnosed as suffering from schizophrenia were no more likely than the rest of the population to have experienced birth complications [4]. Nor were their mothers more likely to have suffered from influenza in the 2nd trimester [5], a finding which acquires cogency from the fact that these mothers were at risk of inflenza in the epidemic that occurred in the autumn of 1957. If the general case for an environmental cause is weak, and if such specific candidates as were presented have been eliminated, one must return to the difficulties of the genetic theory. How can a disease that is genetic have an onset in adult life? Why is it that the majority of cases (perhaps 80%) have no family history of illness? Why is the schizophrenia gene (if such a thing exists) not eliminated from the population if it is associated with such a decrease in fertility?

A Continuum Concept of Psychosis

One of the most characteristic features of psychosis is its age of onset. These disorders rarely manifest themselves before puberty but their incidence then rises steeply (fig. 1). And it rises more steeply in males than in females. The data are taken from the 1945 Survey of Penrose [6] in Ontario, but many others have reported similar ages of onset. The sex difference (that applies particularly to schizophrenia) can be described as the best setablished but least understood finding concerning psychosis.

A number of recent studies indicate that there is more overlap between schizophrenic and affective disorders than is implied by the Kraepelinian concept that these are separate diseases. Kraepelin [7] himself developed doubts about this concept. Thus in 1920 he wrote that

> Perhaps it is ... possible to tackle the difficulties which still prevent us from distinguishing reliably between manic-depressive insanity and dementia praecox. No experienced psychiatrist will deny that there is an alarmingly large number of cases in which it seems impossible, in spite of the most careful observation, to make a firm diagnosis ... Nevertheless it is becoming increasingly clear that we cannot distinguish satisfactorily between these two illnesses and this brings home the suspicion that our formulation of the problem may be incorrect.

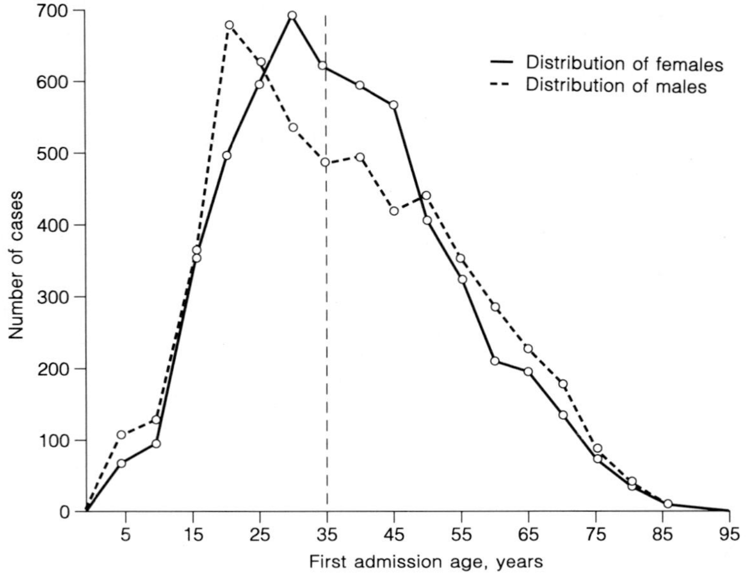

Fig. 1. Age at first hospital admission in relation to sex [from 6].

Two paragraphs later he added

> It is common to find cases of undeniable schizophrenia which show transitory or at times longer-lasting manic-depressive symptoms, and it may be quite impossible to distinguish them from the circular forms of insanity.

These words were written 13 years before Kasanin [8] formulated the concept of schizo-affective psychosis, the ambiguous status of which has been the focus of debate ever since.

Kendell and Gourlay [9] were the first to challenge the Kraepelinian two entities concept in a formal way. In a discriminant function analysis they showed that the symptoms of the two prototypical illnesses schizophrenia and affective psychosis cannot readily be separated into a bimodal distribution. Schizo-affective psychoses, that is illnesses that have both schizophrenic and affective symptoms, are common.

The most compelling evidence is genetic. A number of studies have looked at the question of whether the psychoses breed true within families. For example Odegaard [10] investigated rates of psychotic illness in the first degree relatives of a consecutive series of admissions to the Oslo mental hospitals, and found that as one moves from schizophrenic illness with evidence of a defect state, through lesser severities of schizophrenic and schizoaffective illness to typical manic-depressive disorder the proportion of relatives that received an affective diagnosis increased while the proportion that were diagnosed schizophrenic decreased. While there was a general tendency for relatives to suffer from the same form of illness as probands, at each point on the continuum there was much overlap. Other recent family studies have supported this conclusion. Thus on the basis of their studies in Zurich Angst et al. [11] wrote:

> The obtained results are certainly puzzling in certain aspects. The underlying hypothesis of a continuum is not disproved by our results ... the dichotomy into schizophrenia and affective psychoses is highly questionable ... We do not conclude that a unitary psychosis exists but we think rather of a continuum of psychopathological subgroups with a lot of overlap that may also differ in other respects such as course, genetics and response to treatment.

Angst et al. [11] are surely right to conclude that there is not a unitary psychosis. There are great differences between the psychoses, and Kraepelin was correct to emphasize that there is a relationship between the form of a psychosis and its outcome. In general psychoses in which there is a substantial affective element have a better outcome than those which are more typically schizophrenic in form. But no aetiologically meaningful dividing line can be drawn between them. The existence of the continuum and the variations in form of illness within it are major clues to the origin of psychosis.

The most recent contribution is that of Maier and Lichtermann [12] who studied 525 consecutive hospital admissions with unipolar and bipolar affective, schizoaffective and schizophrenic psychoses and 109 randomly selected controls. 1,250 first-degree relatives were interviewed with the SADS-LA supplemented by interviews for polydiagnostic classification and personality disorders (table 1).

Because the study included a group of control probands one can assess whether the incidence of the various disorders is increased above population expectation. The increase in risk of unipolar affective disorder in the relatives of all categories of psychotic proband is perhaps the salient fea-

Table 1. First-degree relatives' lifetime risk (%) [from 12]

Patient's diagnosis	Schiz	SA	BP	UP
Schizophrenia	5.1	3.3	1.9	19.6
Schizo-affective	4.1	5.3	5.4	20.4
Bipolar affective	1.2	1.0	8.8	23.6
Unipolar affective	0.8	0.9	1.6	21.6
Healthy controls	0.4	0.3	1.2	9.9

ture of the study. By itself it precludes any simple interpretation of the findings. The excess of schizoaffective disorder in the relatives of probands with bipolar illness, and the more modest excess of bipolar disorder in the relatives of patients with schizophrenia and schizoaffective disorder, conflict with any suggestion that a classical Kraepelinian line of demarcation can be drawn between schizo-affective disorders (which thereby are allocated to the category of schizophrenia) and bipolar affective disorder.

What I conclude from these studies is that the findings are more consistent with the concept of a psychotic continuum [13–15] than that there are two distinct diseases as in the original Kraepelinian concept. Here the continuum (represented graphically in fig. 2) extends from unipolar depression through bipolar (manic-depressive) illness to schizoaffective disorder, schizophrenia with and then without significant mood disturbance, and finally to schizophrenia with a defect state (the presence of negative symptoms, e.g. affective flattening or poverty of speech). In a scheme such as this, one can define boundaries which act as markers for particular positions along the continuum. These include the presence of elation, mood-incongruent delusions, nuclear (first-rank or Schneiderian) symptoms, the presence or absence of significant mood disturbance, and negative symptoms. One might ask, what does the continuum represent? An advantage of the concept is that it draws attention to possible correlates of psychosis in the normal population. It seems that the continuum may correspond to variations along a dimension of sociability/emotionality (fig. 3). Thus as one moves from affective disorders toward schizophrenia one moves in a direction of decreasing emotionality and sociability. There is some evidence that relatives of patients with psychoses have non-psychotic features which are related to such characteristics in the probands. Also of interest is the fact that as one moves across the continuum from

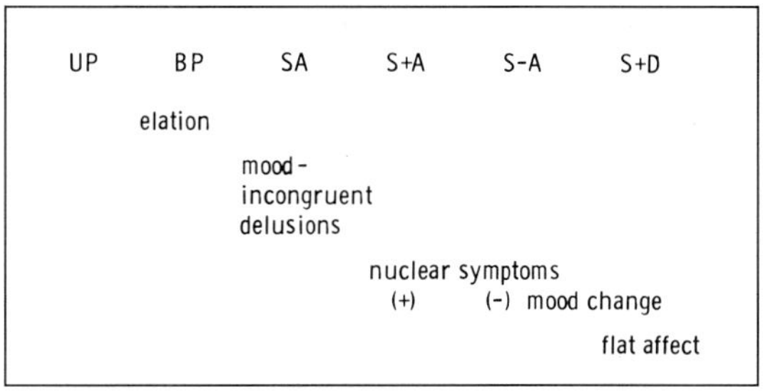

2

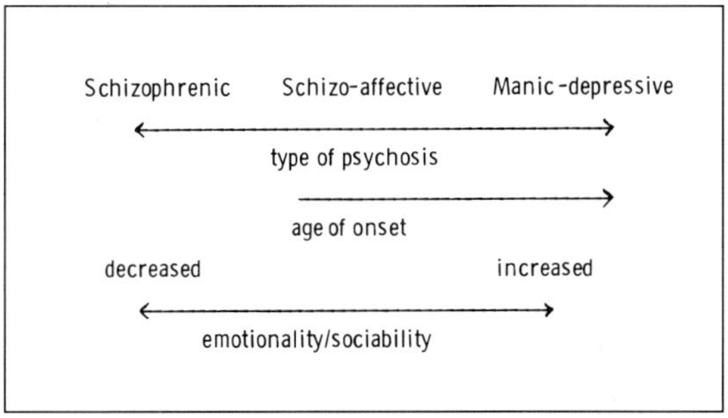

3

Fig. 2. A continuum concept of psychotic illness. UP = Unipolar affective illness; BP = bipolar affective illness; SA = schizo-affective psychosis; S = schizophrenia with and without significant mood disturbance (+ and − A), and with defect state (S + D).

Fig. 3. Possible relationship between the continuum of psychosis and variation in sociability/emotionality in the normal population.

schizophrenia with defect toward affective disorder the mean age of onset is progressively increasing. The continuum can be seen as a lifetime dimension of intellectual and personality development.

Does Form of Illness Predict Response to Treatment?

With the introduction of neuroleptic medication and the demonstration of its efficacy in treatment of schizophrenic illnesses the Kraepelinian distinction acquired apparent therapeutic validity that Kraepelin himself cannot have anticipated. Both in the treatment of acute episodes and in prevention of relapse current practice assumes that the distinction between schizophrenic and affective illnesses has practical relevance. Schizo-

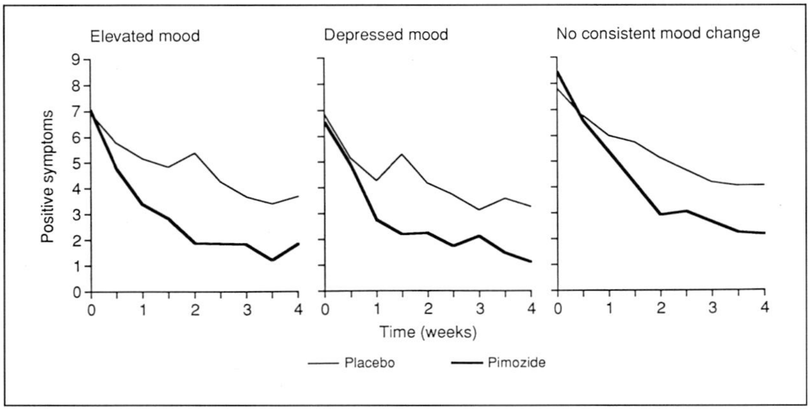

Fig. 4. Effects of pimozide in reducing psychotic symptoms in patients with elevated and depressed mood and without significant mood change [from 16].

phrenic psychoses respond to neuroleptic medication, which also prevents recurrence. Affective psychoses, though they may sometimes be treated with neuroleptics, respond more specifically to lithium. These assumptions are challenged by the findings of the Northwick Park 'functional psychosis' [16, 17]. Patients were included on the basis of the presence of psychotic symptoms (in the absence of an identifiable 'organic' cause). They were not otherwise allocated to a diagnostic category although they were stratified into groups with predominantly elevated, and predominantly depressed mood, and no consistent mood change. The 120 patients entered into the trial were randomly allocated to pimozide (a dopamine antagonist neuroleptic), lithium, a combination of the two medications or placebo.

Pimozide reduced psychotic symptoms in all groups of patients, the effect being as great in patients with predominantly elevated and predominantly depressed moods as in the group without consistent mood change (fig. 4). The only significant effect of lithium was to reduce elevated mood. Thus it appears that the effects of neuroleptic medication (as exemplified by pimozide) are relevant to the resolution of psychotic symptoms in all types of 'functional' psychotic illness. These compounds are antipsychotic rather that antischizophrenic, while the lesser effects of lithium may be limited to mood disturbance.

The benefits of pimozide extended into the follow-up phase of the trial [17]. Amongst patients who had recovered on trial medication those who continued on pimozide fared significantly better ($p = 0.01$) than those who continued on placebo, whereas the benefits of continuation on lithium were not significant. The findings give no support to the view that the diagnostic distinction between the affective and schizophrenic psychoses is a critical predictor of response to neuroleptic treatment, and raise the question of whether there is a wider role for such medication in prophylaxis of affective illnesses than is implied by current practice.

The Meaning of the Brain Changes in Schizophrenia

Brain changes in schizophrenia provide a clue to the nature of the genetic mechanisms underlying the psychotic continuum. Fifteen years ago in the first CT scan study of schizophrenia we reported that some chronic patients have a degree of ventricular enlargement [18]. This finding has been widely replicated although the meaning of the changes has remained unclear. Some recent studies clarify the situation. In the comparison between groups in our study we failed to emphasize an important aspect. Not only was mean ventricular size increased in the patients relative to age-matched controls, but the entire distribution in the patient group was elevated. This important finding has recently been further documented by Harvey et al. [19] who performed an admixture analysis of the ventricle-to-brain ratio (VBR) on computed tomographic (CT) scans from 72 patients with psychosis and 50 healthy controls. The patients showed a more positively skewed distribution and a significantly higher mean VBR, but within the patient group there was no evidence of more than one underlying normal distribution of values. An earlier CT scan study [20] also reached the conclusion that there was no evidence of bimodality in the distribution of values within the group of 136 patients with schizophrenia. In a review of the literature Daniel et al. [21] found no evidence for a mixture of Gaussian distributions within the group of patients with schizophrenia. They concluded that the data did not support the notion that ventricular enlargement is a discontinuous marker for a subtype of schizophrenia.

Thus the morphological brain changes in schizophrenia are not confined to a subgroup of patients but are characteristic of the illness as a whole. In some way the disease process is bound up with brain size or shape. If the disease is genetic the relevant genes are controlling brain development. If affective disorders and schizophrenia are part of the same spectrum then variants of the same developmental genes, and similar or related morphological changes, may also be expected in association with affective disorder.

Asymmetry of Brain Changes

Some studies suggest a more specific role for these genes. Crichton-Browne was a contemporary of Kraepelin who was superintendent of the West Riding Asylum in Yorkshire. In 1879, 12 years after the publication of Darwin's book The Descent of Man, Crichton-Browne [22] published one of the first accounts of a post-mortem study of mental illness. In the introduction he wrote:

> It seemed not improbable that the cortical centres which are last organised, which are the most highly evolved and voluntary, and which are supposed to be located on the left side of the brain, might suffer first in insanity.

There is now a body of evidence from morphological studies of the brain in schizophrenia that suggests that this hypothesis is correct [23]. In our own post-mortem study [24] the components of the lateral ventricle were assessed on radiographs from the lateral aspect after it had been filled with radio-opaque material. Relative to age-matched controls the area of the occipital and temporal horns was increased. The selectivity of the change to the temporal horn confirmed the findings of an earlier study [25]. Most striking in this study was the predilection of the change to the left side (fig. 5). Temporal horn enlargement was also present in Alzheimer-type dementia but in this disease the change was present on both sides. The fact that in schizophrenia the change is confined to the left suggests rather strongly that the disease process is in some way related to the mechanism that determines the asymmetries in the human brain.

Rather similar findings were reported in a magnetic resonance imaging study of 34 patients with first episodes of schizophrenia and 25 controls by Bogerts et al. [26] who concluded that abnormal medial temporal lobe morphology was present in a sub-group of patients at the onset of illness:

> There were clear laterality effects and sex differences: hippocampal tissue was significantly smaller only in the left hemisphere of male patients, whereas enlargement of the whole temporal horn or its anterior portion was present on the left side in both sexes.

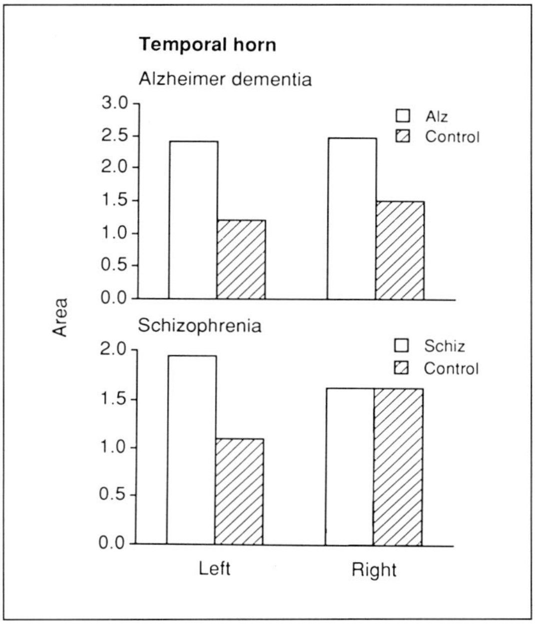

Fig. 5. Temporal horn enlargement in schizophrenia and Alzheimer-type dementia compared to age-matched controls [from 24].

In 1915 Southard [27] summarised his own studies of the pathology of schizophrenia as showing that 'the atrophies and aplasias when focal show a tendency to occur in the left cerebral hemisphere.' He added that 'Aside from the left-sidedness of the lesions very striking is the preference of these changes to occupy the association centers of Flechsig' and 'for this there is probably good a priori reason in the structure, late evolutionary development, and consequent relatively high lability of these regions'. This hypothesis echoes that of Crichton-Browne 25 years earlier but adds emphasis to the notion that it is because these regions have evolved most recently, and because they are variable between individuals, that they are associated with psychotic illness.

Further information concerning the location of the critical changes is available from what is perhaps the most incisive morphological study of schizophrenia yet conducted – the NIMH study of discordant monozygotic twins [28]. An important finding was that relative to the well twin the brain

Table 2. MZ twin pairs discordant for schizophrenia [from 28]

		Left	Right
Anterior temporal grey matter	Well twin	7.86 ± 1.86	8.47 ± 1.27
	Ill twin	7.61 ± 1.36	8.57 ± 1.33
		NS	NS
Total temporal, grey matter	Well twin	28.43 ± 5.99	30.53 ± 6.24
	Ill twin	26.77 ± 5.37	30.21 ± 4.97
		$p < 0.002$	NS

of the ill twin was systematically abnormal in 14 of 15 pairs. This indicates a greater degree of homogeneity of the disease process than might have been expected from studies of singletons. In addition the focus of these changes was in the temporal lobes. Of particular interest in relation to laterality was the finding that there is a reduction in the ill twin in the total volume of grey matter in the temporal lobes on the left but not on the right side (table 2). In that no difference between the sides was present in measures of grey matter in the anterior parts of the temporal lobes the findings suggest rather strongly that there is a structure in the posterior part of the lobe carries a changes in schizophrenia that is lateralised.

The obvious candidate is the planum temporale, the structure that according to Geschwind and Levitsky [29] has the most prominent asymmetry in the human brain (fig. 6). This structure includes auditory association cortex or Wernicke's area, the region that is most concerned with the reception and interpretation of speech. It is of course connected with many other regions on both sides of the brain. If there are changes here in schizophrenia this could account for the axymmetry of the change in the temporal horn and for the fact that some changes are bilateral and include for example enlargement of the third ventricle. It is possible that the morphological changes in the brain in schizophrenia and perhaps in other forms of psychosis can be accounted for as anomalies of development of the planum temporale and its connexions.

Perhaps the most direct evidence of involvement of the planum temporale and cerebral asymmetry comes from post-morten studies of the length of the lateral sulcus. Falkai et al. [30] found a 16% reduction in length of the lateral sulcus in schizophrenic patients on the left side and a loss of asymmetry of the planum temporale. Crow et al. [3] found a similar

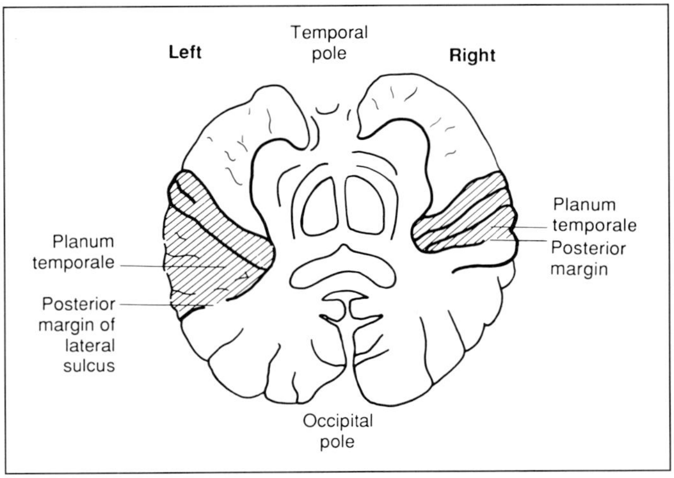

Fig. 6. The major asymmetry in the uman brain in the region of the planum temporale (auditory association cortex, Wernicke's area). The brain is shown as cut along the lateral sulcus to expose the superior surface of the temporal lobe [adapted from 29].

loss of asymmetry of the lateral sulcus in patients in the Runwell 2 series, who earlier had been found to have enlargement of the temporal horn on the left side [24].

Evolution of Language

Why should the anatomy of cerebral asymmetry be so important? The likely explanation is that cerebral asymmetry is a late evolutionary change and one that has been critical in the development of the human brain. Jerison [32] draws attention to the problem of explaining the rapid evolution of the brain in the hominid species. As body weight increases so also does brain weight. Within the species of great apes the increase in brain weight conforms to expectation, being greater in the gorilla than in the chimpanzee and orang-outang. However at certain points in evolution – as for example between the baboon and the great apes there is an increase in brain weight that cannot be accounted for in terms of a simple increase in body size. Jerison refers to this as 'added neural capacity' and the implication is that the brain has acquired new and more complex functions.

In the evolution from australopithecus through homo habilis and homo erectus to homo sapiens there has been an escalation of brain weight in relation to body weight out of all proportion to what had gone before. And these changes have taken place in an evolutionarily short period of time – probably less than 2 million years.

What could have accounted for these changes? According to Jerison [32, p. 405],

> We need not look far for special selection pressures toward the major enlargements of the brain within the genus Homo. The socialized life of a predacious primate is so obviously benefited by linguistic skills, and language is so manifestly the peculiarly human development, that changes in the brain to permit that advantageous supplement to perception and communication would have had obvious selective advantages throughout the period of hominid evolution.

Darwin [33, vol 2, p. 390f.] had also reached this conclusion. According to him

> A great stride in the development of the intellect will have followed, as soon as, through a previous considerable advance, the half-art and half-instinct of language came into use; for the continued use of language will have reacted on the brain, and produced an inherited effect; and this again will have reacted on the improvement of language. The large size of the brain in man, in comparison with that of the lower animals ... may be attributed in chief part ... to the early use of some simple form of language, – that wonderful engine which affixes signs to all sorts of objects and qualities, and excites trains of thought which would never arise from the mere impression of the senses, and if they did could not be followed out.

Deficits in the Development and Structure of Language in Psychosis

There is some evidence that disturbances of language function are present early in schizophrenia and play a central role. Morice and Ingram [34] found that the complexity of language structure was reduced, particularly in patients with an early onset. Thomas et al. [35] reported reduced language complexity in patients with negative symptoms, and a decrease in sentence complexity has been recorded with disease progression [36]. In a retrospective study [37] a group of 100 consecutive patients with first episodes of schizophrenia, schizophreniform and schizo-affective disorder were found to have delayed acquisition of reading skills, and in a prospective study [38], more speech and reading difficulties were noted at the ages

Table 3. Intellectual deficits and psychosis [from 39]

	n	IQ	Age years	p
Male probands	51	88	17	
Their male siblings	44	96	18	<0.01
Their female siblings	41	94	14	<0.001
Their controls	51	101	12	<0.001
Female probands	65	97	13	
Their male siblings	65	96	12	NS
Their female siblings	58	97	16	NS
Their controls	65	102	15	<0.05

of 7, 11 and 16 years in individuals who later developed schizophrenia than in those who suffered from affective disorder or controls.

There is other evidence of intellectual deficits in relation to psychosis. For example it has for some time been known that some individuals who later suffer from schizophrenia do less well on intelligence tests than their well siblings and normal controls long before the onset of illness.

Lane and Albee [40] found deficits in preschizophrenic individuals relative to their siblings at the ages of 7, 11 and 13 years, and Offord [39] found deficits in IQ assessed at school that were more marked in males than in females (table 3). This latter finding is of particular interest in view of the sex difference in age of onset. It is also established that intellectual impairments accumulate in the course of the disease, and that some patients have defects that constitute a true dementia praecox [14]. However there are also indications that onset of illness can be preceded by a period of enhanced intellectual performance. This has sometimes been claimed in relation to schizophrenia [41, 42], but is better documented in the affective disorders. Mason [43] examined the IQ performances of individuals who were subsequently discharged from the US Army on psychiatric grounds and compared them with the general population of army recruits (table 4).

Whereas inductees who later became schizophrenic did less well than the general population – consistent with the findings of Lane and Albee [40] and Offord [39], those who were destined to develop an affective illness had achieved a superior performance. Thus it appears that the genes

Table 4. IQ performances in schizophrenics and manic-depressives [from 43]

	n	p	
All schizophrenics	368	<0.02	low scores
Manic-depressives	25	<0.001	high scores

Table 5. Percentage rates of membership in three different classes of achievement, comparing first-degree relatives of patients with psychosis with the general population, both born 1851–1940 [from 44]

Measure	First-degree relatives (sample 9,379)		General population (sample 171,000)	
	n	%	n	%
Listing in Who's Who	391	4.2	5,813	3.4
Book authorship	184	2.0	2,269	1.3
College graduation	234	2.5	3,485	2.0

that predispose to affective and schizophrenic illness also are related to the development of intellectual function. Perhaps it is a question of timing. Perhaps the genes that are concerned with the optimal development of language function are subject to variation between individuals, and some of those variations result, at a particular point in development, in changes that fall outside the range that fits that individual to his social environment.

Creativity and Achievement in Relation to Psychotic Illness

There is other evidence that the genes that predispose to psychosis are associated with social advantage. Karlsson [44] found that when compared with the general population of Iceland the first-degree relatives of patients admitted to the Reykjavik mental hospital were more likely to have achieved success in being listed in the Icelandic Who's Who, to have written a book, and to have graduated from college (table 5).

Table 6. Age-standardised first admissions for Norway, 1926–1950 in relation to occupation [from 45]

	All diagnoses	Schizophrenia	Manic-depressive illness
General population	690	357	73
Professional service	704	305	151

Thus a genetic relationship to an individual who suffers from psychotic illness may carry social benefits. One may assume that in many of these cases the proband himself suffered from a schizophrenic type illness and that his own social success will have been limited.

Evidence for a more direct benefit of the affective variant of the psychosis gene was accumulated by Noreik and Odegaard (table 6) [45]. Those who achieved a professional level of occupation (teaching, liberal professions, higher administrative posts) were neither more nor less likely than the general population to be admitted to mental hospital. However they were less likely to be admitted with a diagnosis of schizophrenia. This is to be expected when we consider that schizophrenia has an early age of onset and is often crippling with respect to occupational attainment. By contrast they were twice as likely to be admitted with a diagnosis of manic-depressive illness as the general population. Such findings suggest that the affective form of the psychosis gene has benefits with respect to occupational attainment in contemporary society; they add to the body of data accumulated by Jamison [46] suggesting an association between creativity, achievement and affective illness.

Where Is the Cerebral Dominance Gene?

A clue to the location of the cerebral dominance gene comes from the studies of the neuropsychology of Turner's and Klinefelter's syndromes that have been conducted by Netley and Rovet (table 7) [47, 48]. Turner's syndrome cases lack an X chromosome and Klinefelter's have an extra X. Netley and Rovet found that on IQ tests they have reciprocal deficits – in Turner's there is a performance loss [46] and in Klinefelter's a verbal loss [47]. The obvious interpretation is that in Turner's there is a right hemi-

Table 7. Neuropsychology of Turner's and Klinefelter's syndromes

	Chromosomes	n	Verbal IQ	Performance IQ
Turner's syndrome	XO	35	100	88
Klinefelter's syndrome	XXY	24	83	100

Note: In a collated sample of XXX girls a verbal-performance discrepancy similar to, but of lesser magnitude than, that in XXY males was seen [48].

Table 8. Siblings both suffering from psychosis [from 6]

	Brother-brother	Brother-sister	Sister-sister
Schizophrenia	106	192	124
Affective disorder	69	140	118

sphere deficit and in Klinefelter's a left hemisphere deficit. Thus some factor located on the X chromosome is responsible for the relative development of the two hemispheres. However because in normal females one X chromosome is inactivated the abnormalities in Turner's syndrome are attributable to those parts of the X that are not inactivated. This suggests that the cerebral dominance gene is in the pseudoautosomal region or some other part of the X chromosome that is not inactivated [50].

Where Is the Psychosis Gene?

In his 1945 survey of familial mental illness in Ontario, Penrose [6] reported that sibling pairs with both affective disorder and schizophrenia were more likely to be of the same sex (table 8). Same sex concordance in schizophrenia was discussed by Penrose [51] and Rosenthal [52]. Penrose [6] also drew attention to the fact that age of onset correlations between pairs of affected relatives are consistent with X linkage. X linkage although sometimes considered for affective illness is discounted for psychosis in general on the grounds that father-to-son transmission is not infrequent. However there are two ways in which a locus on the sex chromosomes is compatible with an autosomal pattern of transmission: (1) that the locus is within the pseudoautosomal region [53], the telomeric 2.6mb of the short arms of the X and Y chromosomes within which there is a single obligatory recombination in male meiosis, or (2) that the gene is represented in homologous form in the non-recombining portions of the X and Y chromosomes. The defining characteristic of X and Y linkage (whether pseudoautosomal or associated with an X-Y homologous gene) is same sex concordance – within families affected individuals above chance expectation will be of the same sex. This arises because when such a gene is inherited from a father it will be preferentially passed either on the X chromosome to daughters or on the Y chromosome to sons. By contrast when the disease is inherited on an X chromosome from the mother it is equally likely to be passed to a daughter or a son. The prediction that concordance by sex for psychosis will apply particularly to those cases in which inheritance is through the father has been tested in a series of 120 sibships that include at least two psychotic members [54].

Of the two possible forms of sex linkage outlined above (a pseudoautosomal gene, and an X-Y homologous gene) the latter is favoured by the following considerations: (1) In the Penrose 1945 survey [6] 28 father-son pairs with schizophrenia were observed compared to 16 father-daughter pairs. An excess of father-son relative to father-daughter pairs has been observed in some other studies [24] and is compatible with Y linkage. (2) Because typically, due to lack of recombination, the gene sequences on the X and Y chromosome diverge, an X-Y homologous gene provides a possible explanation for a sex difference, e.g. with respect to age of onset.

Darwin [33] distinguished between 'natural' and 'sexual' selection and attributed to the latter those characteristics which are acquired by members of one sex (commonly males) as a result of inter-male competition or mate choice by females. Characters acquired in this way might later be transmitted to the other sex. One way in which this might happen, it seems, it by mutations on the sex chromosomes. According to Darwin sexual selection has been relevant to cerebral development in man. Thus,

> He who admits the principle of sexual selection will be led to the remarkable conclusion that the cerebral system not only regulates most of the existing functions of the body, but has indirectly influenced the progressive development of various bodily structures and mental qualities. Courage, pugnacity ... musical organs both vocal and instrumental ... have all been gained by the one sex or the other, through the influence of love and jealousy, through the appreciation of the beautiful in sound, colour or form, and through the exertion of a choice ... [p. 402].

For such reasons it is plausible that language and other higher functions in man have evolved as a result of changes in the sex chromosomes, and particularly in relation to the interaction between X and Y. If these lately evolved functions also relate, as Darwin suggested, to intraspecies individual differences, the underlying genetic mechanisms are presumably subject to a high degree of variation. This may be why they are associated with high rates of pathology.

Conclusions

(1) Psychotic illnesses are common and variable in form, but occur in all societies, and are associated with a fertility disadvantage. If as seems likely these illnesses are genetic in origin, some explanation of their persistence is called for. (2) Brain changes (e.g. ventricular enlargement) in schizophrenia are consistent (i.e. not confined to a subgroup) and explicable on the basis that there is an arrest of development of asymmetry (most easily detected in the left lateral sulcus) associated with more generalised and perhaps secondary changes elsewhere. (3) When viewed as a whole the psychoses can be seen to be associated with variations in intellect and language function in both probands and relatives that suggest that the underlying genetic mechanisms have played an important role in the evolution of language and social intelligence in man. (4) The hypothesis is proposed that these evolutionary developments have been brought about by a process of sexual selection in which an X-Y homologous gene (the cerebral dominance gene) that is responsible for the relative rates of development of the two cerebral hemispheres, has played a critical role.

References

1 Jablensky A, Sartorius N, Ernberg G, Anker M, Korten A, Cooper JE, Day R, Bertelsen A: Schizophrenia: Manifestations, incidence and course in different cultures. A World Health Organization ten-country study. Psychol Med 1992;suppl 20.

2 Crow TJ, Done DJ: Age of onset of schizophrenia in siblings: A test of the contagion hypothesis. Psychiatry Res 1986;18:107–117.

3 Karlsson J: The rate of schizophrenia in foster-reared close relatives of schizophrenic index cases. Biol Psychiatry 1970;2:285–290.

4 Done DJ, Johnstone EC, Frith CD, Golding J, Shepherd PM, Crow TJ: Complications of pregnancy and delivery in relation to psychosis in adult life: A study using the British perinatal mortality survey. Br Med J 1991;302:1576–1580.

5 Crow TJ, Done DJ, Johnstone EC: Schizophrenia and influenza. Lancet 1991;338: 116–117,
6 Penrose LS: Survey of cases of familial mental illness. Eur Arch Psychiatry Clin Neurosci 1991;240:315–324.
7 Kraepelin E: Die Erscheinungsformen des Irreseins. Z Ges Neurol Psychiat 1920;62: 1–29; translated by H Marshall as: Patterns of mental disorder; in Hirsch SR, Shepherd M (eds): Themes and Variations in European Psychiatry. Bristol, Wright, 1974, pp 7–30.
8 Kasanin J: The acute schizo-affective psychoses. Am J Psychiatry 1933;90:97–126.
9 Kendell RE, Gourlay J: The clinical distinction between the affective-psychoses and schizophrenia. Br J Psychiatry 1970;117:261–266.
10 Odegaard O: The multifactorial inheritance of predisposition to schizophrenia; in Kaplan AR (ed): Genetic Factors in 'Schizophrenia'. Springfield, Thomas, 1972, pp 256–275.
11 Angst J, Scharfetter C, Stassen HH: Classification of schizo-affective patients by multidimensional scaling and cluster analysis. Psychiatr Clin 1983;16:254–264.
12 Maier W, Lichtermann D: The interrelationship of major psychotic disorder as evidenced by patterns of familial transmission; in Racagni G, Brunello N, Fukuda T eds: Biological Psychiatry. Amsterdam, Elsevier, 1991, vol 2, pp 503–505.
13 Crow TJ: The continuum of psychosis and its implication for the structure of the gene. Br J Psychiatry 1986;149:419–429.
14 Crow TJ: Temporal disorientation in chronic schizophrenia: The implications of an 'organic' psychological impairment for the concept of 'functional' psychosis; in Kerr A, Snaith P (eds): Contemporary Issues in Schizophrenia. London, Gaskell, 1986, pp 168–174.
15 Crow TJ: Nature of the genetic contribution to psychotic illness: A continuum viewpoint. Acta Psychiat Scand 1990;81:401–408.
16 Johnstone EC, Crow TJ, Frith CD, Owens DGC: The Northwick Park 'functional' psychosis study: Diagnosis and treatment response. Lancet 1988;ii:119–126.
17 Johnstone EC, Crow TJ, Owens DGC, Frith CD: The Northwick Park 'functional' psychosis study. Phase 2: Maintenance treatment. J Psychopharmacol 1991;5:388–395.
18 Johnstone EC, Crow TJ, Frith CD, Husband J, Kreel L: Cerebral ventricular size and cognitive impairment in chronic schizophrenia. Lancet 1976;ii:924–926.
19 Harvey I, McGuffin P, Williams M, Toone BK: The ventricle-brain ratio (VBR) in functional psychoses: an admixture analysis. Psychiatr Res Neuroimaging 1990;35: 61–69.
20 Owens DGC, Johnstone EC, Crow TJ, Frith CD, Jagoe JR, Kreel L: Cerebral ventricular enlargement in schizophrenia: relationship to the disease process and its clinical correlates. Psychol Med 1985;15:27–41.
21 Daniel DG, Goldberg TE, Gibbons RD, Weinberger DR: Lack of a bimodal distribution of ventricular size in schizophrenia: A Gaussian mixture analysis of 1056 cases and controls. Biol Psychiatry 1991;30:887–903.
22 Crichton-Browne J: On the weight of the brain and its component parts in the insane. Brain 1879;2:42–67.

23 Crow TJ: Temporal lobe asymmetries as the key to the aetiology of schizophrenia. Schizophr Bull 1990;16:433–443.

24 Crow TJ, Ball J, Bloom SR, Brown R, Bruton CJ, Colter N, Frith CD, Johnstone EC, Owens DGC, Roberts GW: Schizophrenia as an anomaly of the development of cerebral asymmetry. Arch Gen Psychiatry 1989;46:1145–1150.

25 Brown R, Colter N, Corsellis JAN, Crow TJ, Frith CD, Jagoe R, Johnstone EC, Marsh L: Post-mortem evidence of structural brain changes in schizophrenia. Arch Gen Psychiatry 1986;43:36–42.

26 Bogerts B, Ashtari M, Degreef G, Alvir JMJ, Bilder RM, Lieberman J: Reduced temporal limbic structure volumes on magnetic resonance images in first episode schizophrenia. Psychiatry Res Neuroimaging 1990;35:1–13.

27 Southard EE: On the topographical distribution of cortex lesions and anomalies in dementia praecox, with some account of their functional significance. Am J Insanity 1915;71:603–671.

28 Suddath RL, Christison GW, Torrey EF, Casanova MF, Weinberger DR: Anatomical abnormalities in the brains of monozygotic twins discordant for schizophrenia. N Engl J Med 1990;322:789–794.

29 Geschwind N, Levitsky W: Left-right asymmetry in temporal speech region. Science 1968;161:186–187.

30 Falkai P, Bogerts B, Greve B, Folsch-Reetz B, Pilz K, Gonsiorczyk C, Pfeiffer U, Machus B: Reduced Sylvian fissure and planum temporale asymmetry in schizophrenia: Evidence for disturbed left hemispheric neurodevelopment? Schizophr Res 1992;6:152.

31 Crow TJ, Brown R, Bruton CJ, Frith CD, Gray V: Loss of Sylvian fissure asymmetry in schizophrenia: Findings in the Runwell 2 series of brains. Schizophr Res 1982;6: 152–153.

32 Jerison HJ: Evolution of the Brain and Intelligence. New York, Academic Press, 1973.

33 Darwin C: The Descent of Man, and Selection in Relation to Sex. Princeton, 1981, vol 2, pp 390–391/402.

34 Morice RD, Ingram JCL: Language complexity and age at onset of schizophrenia. Psychiatry Res 1983;9:233–242.

35 Thomas PF, King K, Fraser WI: Positive and negative symptoms of schizophrenia and linguistic performance. Acta Psychiat Scand 1987;76:144–151.

36 King K, Fraser WI, Thomas P, Kendell RE: Re-examination of the language of psychotic subjects. Br J Psychiatry 1990;156:211–215.

37 DeLisi LE, Boccio A, Riordan H, Hoff AL, Dorfman A, McClelland J, Kushner M, van Eyl O, Oden N: Familial thyroid disease and delayed language development in first admission patients with schizophrenia. Psychiatry Res 1991;38:39–50.

38 Done DJ, Crow TJ, Johnstone EC, Sacker A: Childhood antecedents of schizophrenia and affective disorder – a prospective study. Schizophr Res 1992;6:105.

39 Offord DR: School performance of adult schizophrenics, their siblings and age mates. Br J Psychiatry 1974;125:12–19.

40 Lane EA, Albee GW: Early childhood intellectual differences between schizophrenic adults and their siblings. J Abnorm Soc Psychol 1964;68:193–195.

41 Parfitt DN: The neurology of schizophrenia. J Ment Sci 1956;102:671–718.

42 Fitzherbert J: Increase in intelligence quotient at onset of schizophrenia: Three adolescent cases. Br J Med Psychol 1955;28:191–193.
43 Mason CF: Pre-illness intelligence of mental hospital patients. J Consult Psychol 1956;20:297–300.
44 Karlsson JL: Creative intelligence in relatives of mental patients. Hereditas 1984; 100:83–86.
45 Noreik K, Odegaard O: Psychoses in Norwegians with a background of higher education. Br J Psychiatry 1966;112:43–55.
46 Jamison KR: Manic-depressive illness and accomplishment: Creativity, leadership, and social class; in Goodwin FK, Jamison KR (eds): Manic-Depressive Illness. Oxford; Oxford University Press, 1990, p 999.
47 Netley CT, Rovet J: Atypical hemispheric lateralization in Turner syndrome subjects. Cortex 1982;18:377–384.
48 Netley CT, Rovet J: Relations between a dermatoglyphic measure, hemispheric specialization, and intellectual abilities in 47,XXY males. Brain Cognition 1987;6:153–160.
49 Netley CT: Summary overview of behavioural development in individuals with neonatally identified X and Y aneuploidy. Birth Defects 1986;22:293–306.
50 Crow TJ: Pseudoautosomal locus for the cerebral dominance gene. Lancet 1989;ii: 339–340.
51 Penrose LS: Auxiliary genes for determining sex as contributory causes of mental illness. J Ment Sci 1842;88:308–316.
52 Rosenthal D: Familial concordance by sex with respect to schizophrenia. Psychol Bull 1962;59:401–421.
53 Crow TJ: Sex chromosomes and psychosis: The case for a pseudoautosomal locus. Br J Psychiatry 1988;153:675–683.
54 Crow TJ, DeLisi LE, Johstone EC: Concordance by sex in sibling pairs with schizophrenia is paternally inherited: Evidence for a pseudoatosomal locus. Br J Psychiatry 1989;155:92–97.

Timothy J. Crow, Division of Psychiatry, Clinical Research Centre,
Watford Road, Harrow, HA1 3UJ (UK)

Brunello N, Mendlewicz J, Racagni G (eds): New Generation of Antipsychotic Drugs: Novel Mechanisms of Action. Int Acad Biomed Drug Res. Basel, Karger, 1993, vol 4, pp 62–66

Pharmacology of Sertindole, a Novel Atypical Neuroleptic

Torben Skarsfeldt[1]

Pharmacological Research, H. Lundbeck A/S, Copenhagen-Valby, Denmark

Introduction

Since the introduction of the first neuroleptic compound, chlorpromazine, in the mid-1950s, extensive work has been done to develop new compounds that do not induce acute extrapyramidal side effects (EPS; dystonia and parkinsonian-like symptoms). However, almost all of the neuroleptics introduced have been classical (they induce EPS), since only clozapine has been characterized as an atypical (it does not induce EPS). While EPS are believed to result from dopamine (DA) blockade in the nigrostriatal system, the antipsychotic action involves inhibition of the limbic DA system [1]. Sertindole – 1-[2-[4-[5-chloro-1-(4-fluorophenyl)-1H-indol-3-yl]-1-piperidinyl]ethyl]-2-imidazolidinone – is a new putative neuroleptic with extreme selectivity for the limbic DA system after repeated treatment. The present study reviews the pharmacological profile of sertindole in vitro and in vivo, and further shows the selectivity for the limbic DA system after repeated treatment.

Pharmacology in vitro

Sertindole has high affinity (nanomolar range) for 5-HT_2, α_1-adrenergic, and DA D_2-receptors [2]. The affinity for DA D_1-receptors is low but still in the nanomolar range (table 1). The affinity for α_2-adrenoceptors,

[1] I thank the staff of the research laboratories at H. Lundbeck A/S for contributions to this review.

Table 1. Receptor-binding profile of sertindole, clozapine and haloperidol

Binding site	Ligand	Affinity (IC_{50}, n*M*)		
		sertindole	clozapine	haloperidol
$5\text{-}HT_2$[1]	^3H-ketanserin	0.39	7.8	55
$5\text{-}HT_{1A}$[1]	^3H-8-OHDPAT	2,600	860	3,200
$5\text{-}HT_3$	^3H-LY 278584	> 100,000	180	76,000
DA D_1[1]	^3H-SCH 23390	28	130	36
DA D_2[1]	^3H-spiperone	4.1	410	7.5
NA, α_1[1]	^3H-prazosin	3.4	9.2	18
NA, α_2[1]	^3H-idazoxan	350	64	2,000
NA, β[1]	^3H-dihydroalprenolol	39,000	81,000	14,000
Histamine H_1[1]	^3H-mepyramine	600	23	> 1,000
ACh, muscarine[1]	^3H-QNB	2,500	9.4	5,500
Sigma[1]	^3H-DTG	73	1,600	17
Phencyclidine	^3H-TCP	13,000	32,000	4,600

NA = Noradrenaline; IC_{50} = concentration causing 50% inhibition of binding.
[1] Results previously published by Sánchez et al. [2].

histamine H_1-receptors, and sigma receptors is lower, and no affinity for $5\text{-}HT_{1A}$-receptors, muscarine cholinergic receptors, and β-adrenoceptors is seen. The receptor profile of sertindole deviates from the profile of haloperidol, which has higher affinity for DA D_2 than for $5HT_2$-receptors and α_1-adrenoceptors. The receptor binding profile of sertindole also deviates from that of clozapine, which has high affinity for muscarinic cholinergic receptors and low affinity for DA D_2-receptors. Ex vivo binding studies confirm the in vitro profile (effect on $5\text{-}HT_2 > \alpha_1 >$ DA D_2) and maximal inhibition is obtained within the first 48 h after dosing. The effect is still observed 10 days after dosing [3].

In vivo Pharmacology

Sertindole has no DA antagonistic activity in the classical test models for stereotypy (methylphenidate, amphetamine) and does not induce catalepsy. This is in contrast to haloperidol, which potently inhibits stereotypy and induces catalepsy (table 2). Clozapine does not induce catalepsy and

Table 2. Antagonism by sertindole, clozapine and haloperidol of dopaminergic and α_1-adrenergic effects, which are centrally mediated

Test model	Pre-treatment time, h	ED_{50}, μmol/kg, s.c.		
		sertindole	clozapine	haloperidol
DA antagonism				
Methylphenidate gnawing, mice	2[1]	>180	98	0.30
Amphetamine stereotypy, rats	2[1]	>45	>120	0.15
Catalepsy, rats, max. eff.	1–6[1]	>91	>120	0.36
Serotonin antagonism				
Quipazine-induced head twitches	2	0.035	1.5	0.16
	24	0.030	>7.6	>13
	24, p.o.	0.039	>31	>13
d-LSD cue	2	0.17	6.4	>0.42

ED_{50} = Dose producing a desired effect in 50% of the study population.
[1] Results previously published by Sánchez et al. [2].

only weakly inhibits stereotypies. Sertindole is an extremely potent 5-HT_2-receptor antagonist (quipazine-induced head twitches) and the duration of action is very long (table 2). After a single injection of sertindole (2.9 μmol/kg, s.c. = 1.3 mg/kg) a maximum effect (100% inhibition of head twitches) is observed during the first 3 days and even after 9 days some 30% inhibition is apparent [2]. In acute electrophysiological experiments intravenous administration of sertindole – in contrast to both clozapine and haloperidol – reverses neither *d*-amphetamine- nor apomorphine-induced inhibition of the firing frequencies of DA neurons in substantia nigra pars compacta (SNC) or in ventral tegmental area (VTA) [4]. The acute intravenous administration of sertindole does not influence the firing frequencies of the spontaneously active DA neurons either in SNC or in VTA. However, 3 weeks' treatment with sertindole selectively inhibits the number of spontaneously active DA neurons in VTA, while leaving the number of active DA neurons in SNC unaffected. The selectivity ratio between the potencies in VTA and SNC is higher than 100 for sertindole, which is much higher than that observed with other neuroleptics (table 3). Acute injection of baclofen or apomorphine reverses the inhibi-

Table 3. Effect of neuroleptics on DA neurons in VTA and SNC after 3 weeks of treatment

Compound	ED_{50}, µmol/kg/day, p.o.		Ratio SNC/VTA
	VTA	SNC	
Sertindole	0.015	1.6	107
Clozapine	56	> 120	> 2.1
Haloperidol	0.043	0.033	0.77

ED_{50} = Dose producing a desired effect in 50% of the study population.

tion of activity seen after repeated treatment with sertindole. This suggests that sertindole induces depolarization inactivation of the DA neurons. The inactivation is reversible since normal activity of DA neurons is found in both SNC and VTA 2 weeks after withdrawal from repeated treatment with a low dose of sertindole. Depolarization inactivation induced by sertindole requires 3 weeks of treatment as has been reported for other neuroleptics.

Conclusion

In conclusion, sertindole has unique pharmacological properties. Sertindole has a low affinity for muscarinic receptors and only shows a weak antagonistic effect on peripheral α_1-adrenoceptors [data not shown], making autonomic side effects unlikely. Sertindole has a favorable ratio between activity after parenteral and oral administration. The in vivo profile of sertindole is atypical. In all acute tests for antidopaminergic activity, sertindole is remarkably weak or without effect. However, sertindole shows extreme potent and long-acting 5-HT_2-receptor antagonism. The marked neurophysiological selectivity for DA neurons in VTA as compared with the effect on the DA neurons in SNC after repeated treatment indicates a large ratio between doses inducing antipsychotic effect and those inducing EPS. Therefore, sertindole is expected to be an atypical neuroleptic compound, effective in the treatment of schizophrenic disorders without the potential for EPS.

References

1 Chiodo LA, Bunney BS: Typical and atypical neuroleptics: Different effects of chronic administration on the activity of A9 and A10 midbrain dopaminergic neurons. J Neurosci 1983;3:1607–1619.

2 Sánchez C, Arnt J, Dragsted N, Hyttel J, Lembøl HL, Meier E, Perregaard J, Skarsfeldt T: Neurochemical and in vivo pharmacological profile of sertindole, a limbic-selective neuroleptic compound. Drug Dev Res 1991;22:239–250.

3 Hyttel J: Sertindole – a new concept in neuroleptic research. Biol Psychiatry 1991; 29:33S:S10–15-01.

4 Skarsfeldt T: Electrophysiological profile of the new atypical neuroleptic, sertindole, on midbrain dopamine neurones in rats: Acute and repeated treatment. Synapse 1992;10:25–33.

Torben Skarsfeldt, MSc, Pharmacological Research, H. Lundbeck A/S, Ottiliavej 9, DK–2500 Copenhagen-Valby (Denmark)

Brunello N, Mendlewicz J, Racagni G (eds): New Generation of Antipsychotic Drugs: Novel Mechanisms of Action. Int Acad Biomed Drug Res. Basel, Karger, 1993, vol 4, pp 67–76

Nontarget Actions of Neuroleptics

Neuroleptic Side Effects and Tardive Dyskinesia

Carol A. Tamminga[a], *Osamu Shirakawa*[b], *Robert R. Conley*[a]

[a] University of Maryland, Maryland Psychiatric Research Center, Baltimore, Md., USA; [b] Kobe University School of Medicine, Department of Psychiatry and Neurology, Kobe, Japan

Introduction

The main target action of neuroleptics is their antipsychotic effect; other actions are usually considered side effects. These other effects are interesting in relationship to one's perspective: patients do not like them; doctors try to minimize them; pharmaceutical houses develop drugs to avoid them, and neurochemists attempt to study them to yield new information. Neuroleptics have many different kinds of actions, due to their primary antidopaminergic activity, their broad monoaminergic receptor blockade and the critical regulatory role these transmitter systems play in mammalian behavior. Curiously, not only do neuroleptics have acute nontarget actions, like most drugs, but they also have delayed-onset, or tardive involuntary motor effects. The onset and offset of these tardive actions are extended over months and years, thus making the nature of these side effects nontrivial, i.e. patients cannot rid themselves of the effects simply by stopping drug. Hypotheses of the neural mechanisms determining tardive dyskinesia are interesting, and perhaps unique to drug action in brain.

Acute Neuroleptic Side Effects

Acute nontarget neuroleptic effects are widely known and have been previously fully described (table 1). Simply stated, the earliest neuroleptics were phenothiazines, with broad monoaminergic and muscarinic actions

Table 1. Acute neuroleptic side effects

Side effect	Mechanism	Associated neuroleptic
Parkinsonism	DA blockade	Haloperidol
Dystonia	DA blockade	Haloperidol
Akathesia	DA blockade	Haloperidol
Hypotension	Muscarinic, α_2 blockade	Chlorpromazine
Sedation	Muscarinic, $5HT_2$ blockade	Chlorpromazine
Tachycardia	Muscarinic blockade	Chlorpromazine
Seizures	Unknown	All neuroleptics
Rash	Peripheral deposition	Phenothiazines
Corneal/lens opacities	Peripheral deposition	Phenothiazines
Skin pigmentation	Peripheral deposition	Phenothiazines
Retinitis	Peripheral deposition	Phenothiazines
Jaundice	Co-precipitates in bile ducts	All neuroleptics
Agranulocytis	Attacks mature granulocytes	All neuroleptics
QRS changes	Muscarinic blockade	All neuroleptics

and low antipsychotic potency. Chlorpromazine is the classic example, as well as being the first neuroleptic [1]. Phenothiazine side effects are predominantly sedation, tachycardia, and hypotension. Acute motor reactions, like parkinsonism, dystonias, and akathisia, occur with phenothiazines, although to a lesser degree than with butyrophenones. The low potency of the phenothiazines provides a wide dose window between the onset of antipsychotic efficacy and motor side effects. Nonphenothiazine antipsychotics include the dibenzoxazepine, butyrophenones, thioxanthenes, diboroxazepine (dibenzodiazepine), and benzamide groups. In general, these drugs have lower cardiac and other peripheral side effects and more pronounced acute motor reactions. The dose window between antipsychotic efficacy and motor side effect onset is slight.

One tract of drug development in the area of antipsychotics has been to make drugs with reduced acute nondopaminergic side effects; this has been done chiefly by purifying the broad action antimonoaminergic effects and focusing action on antagonism of D_2-receptors. Previous pharmacologic study has suggested that the mechanism of antipsychotic effect is D_2

antagonism [2, 3]. This may now be more complex, however, because multiple D_2 family dopamine (DA) receptor subtypes are known to exist and are likely to have distinct functions. Nonetheless, drugs have been developed with relatively pure D_2 antagonism, relative to other transmitter receptors. These drugs retain motor side effects, but have shed many of the other cardiovascular effects. With the discovery and cloning of multiple DA receptors, and the possibility that each may be functionally distinct, a refinement in this direction of drug development is possible. Predominant action at one or another of the DA receptor subtypes may maximize antipsychotic action and minimize side effects. The current DA receptors still fall into two families, the D_1 and the D_2 groups. No correlation has been demonstrated between D_1 blockade and antipsychotic action. Nonetheless, based on a solid rationale and a potent selective drug, a D_1 antagonist is currently being tested in schizophrenia. As selective antagonists are developed for the D_2 family of receptors, their actions will be evaluated with interest.

The other tract of drug development has been to design clozapine-like antipsychotic drugs, with both their positive antipsychotic effects and potent side effect profile. It has been recognized that clozapine is a drug with unique antipsychotic actions [4]. The exact mechanism of this action is not yet known. Several theories have been put forward, including its combined $D_2/5HT_2$ antagonist action; its balanced D_1/D_2 antagonist action; its low affinity for the striatal D_2 receptor and relatively higher limbic D_2 affinity, in vivo; and perhaps, its overall mixed receptor antagonist profile. While waiting to identify exactly its mechanism of action, drugs have been synthesized to mimic the mixed receptor profile of clozapine. These drugs may well have the same (rather weighty) side effect profile as clozapine, and may hopefully possess some of its unique target actions.

Acute Typical versus Atypical Neuroleptic Side Effects

A mere listing of the side effects of the atypical 'broad-spectrum' antimonoaminergic drugs like clozapine, is daunting [5]. That the drug is used at all attests to its special antipsychotic characteristics. All clinicians have experienced that, aside from the agranulocytosis, clozapine often causes treatment-limiting side effects, such as tachycardia, seizures, QRS changes on EKG, lethargy, drooling, and weight gain. Motor side effects are considerably reduced with clozapine, but still may occur, however mildly. No other 'atypical' antipsychotic drugs have yet been finally identified, so a comparison of side effects is not possible. Efforts to identify such a series of drugs are active.

Antipsychotic drugs have traditionally fallen on a scale that roughly counterbalanced motor and cardiovascular side effects. Haloperidol and sulpiride are both examples of so-called high potency drugs which are notable neurochemically for a relatively specific postsynaptic D_2-receptor blockade. These drugs manifest few cardiovascular side effects, theoretically secondary to selective dopamine action and high milligram antipsychotic potency allowing relatively low doses to be used clinically. The side effect 'cost' has been a high incidence of hypokinetic involuntary symptoms. On the other end of this spectrum, drugs such as chlorpromazine, thioridazine, and clozapine have fewer motor side effects but significantly more cardiovascular problems. These latter drugs are relatively nonspecific monoaminergic postsynaptic receptor blockers. These characteristics lead to the hypothesis that their anticholinergic and possibly antiadrenergic effects have successfully opposed the neuroleptic-induced parkinsonism and other motor effects, at the expense of other kinds of drug side effects. These drugs also have low milligram dose potency and the relatively high doses required for clinical effect also contributes to a large number of adverse effects.

It is important to note that low and high potency typical neuroleptics exhibit the same propensity of causing tardive dyskinesia (TD). This may not be the case with the so-called atypical antipsychotics which are notable for causing few motor symptoms. The prototypic drug of this class, clozapine, is a fairly nonspecific agent in its binding profile, similar to chlorpromazine. One could assume clozapine's low motor effect liability would thus be secondary to its marked antimuscarinic effects. Indeed this may be the case. However, other explanations seem more likely, based on the observation that even very high dose clozapine produces no catalepsy in animals [6]. The selective activity the drug appears to show for the limbic system [7] possiblly through its low affinity to D_2-receptor [8], or the possibility that clozapine could effect another critical receptor as primary site of action are examples of these other possibilities. This leads us to the hypothesis that other side effects mediated in striatal motor areas, such as TD, will be rare with this drug. That this might be the case with clozapine is being tested prospectively and concurrently right now [G.K. Thaker, personal commun.]. Moreover, this hypothesis will be further tested with the study of remoxipride and similar putatively 'limbic selective' specific anti D_2 agents, which, if they demonstrate low motor effects in humans, clearly must do so for reasons other than antimuscarinic activity.

Neuroleptic-Induced Late Onset Dyskinesia

Prolonged treatment of schizophrenia with neuroleptics induces the delayed onset of a hyperkinetic involuntary motor disorder called TD. Its incidence is 4–5% per treatment year for typical neuroleptics [9]. Although it is neither permanent nor progressive, as early speculation had it, it can nonetheless be very serious. Clinical problems range all the way from an unsightly presentation which interferes with independent living, to impaired swallowing and temporal-mandibular joint malfunction. Medical-legal implications are always present because it is an iatrogenic syndrome. Risk factors for developing TD are affective illness or a schizoaffective diagnosis; high dose and prolonged neuroleptic use; and severe schizophrenic symptoms. The study of this syndrome is imperative because of its clinical import and the need to understand and control the neural mechanisms causing these motor sequelae. Although its etiology is chronic neuroleptic treatment, the mechanism of TD is currently not known. Early theories centered around DA receptor upregulation. Specifically, the early theory suggested that the neuroleptic-induced increase in DA receptor number in striatum produced the gradual onset of dyskinetic movements in patients. This hypothesis was consistent with one of the actions of neuroleptic drugs, namely a gradual increase in DA receptor numbers in terminal areas with chronic treatment [10]; and it was also supported by the behavioral pharmacology of the disorders [11]. Subsequent research, however, has shown no selective upregulation of any DA receptors in the neuroleptic-treated brains of TD schizophrenics compared to non-TD patients. No pharmacologic probe tests of dopaminergic function have been able to distinguish TD from non-TD schizophrenia. And the in vivo visualization of DA receptors has failed to distinguish TD from non-TD schizophrenia. Therefore, this hypothesis, by itself, is no longer plausible [12].

Because of the profusion of GABA containing neurons in the basal ganglia, both as interneurons and long tract neurons, a neuroleptic-induced abnormality of the GABAergic system associated with TD has always been an active possibility [12]. Chronic neuroleptic treatment increases GABA receptor number in the substantia nigra, pars reticulata (SNR) [13]. Gunne et al. [14] were first to report that reductions in glutamic acid decarboxylase, the synthetic enzyme for GABA, discriminated between monkeys chronically treated with neuroleptics who did and who did not develop hyperkinetic oral facial movements (i.e. dyskinesia-like). Moreover, clinically, the therapeutic efficacy of GABA agonists in these dyskinesias was identified, hence providing supportive experimental clinical data [12].

Involvement of the GABA system in the pathophysiology of TD has been supported not only by the clinical pharmacology of GABA agonists in TD, but also by several other GABA-related measures which distinguished TD from non-TD schizophrenic patients. First potent and selective GABAmimetic drugs all improve dyskinetic symptoms in TD. Second, TD schizophrenics (free from neuroleptic medication) have lower lumbar cerebrospinal fluid concentrations of GABA than non-TD patients [12]. Third, TD schizophrenic subjects have a high degree of oculomotor 'distractibility' when fixing on a light target, in contrast to most non-TD schizophrenics who behave like normal controls [15]. This ocular 'distractibility' is mediated, at least in nonhuman primates, by reduced GABAergic activity in the nigral GABA-containing efferent tract to the superior colliculus. These characteristics of TD schizophrenics are consistent with the idea of augmented GABAergic transmission at GABA-mediated synapses in the SNR.

In order to facilitate a detailed study of GABAergic mechanisms in tardive dyskinesia, we have developed a rodent model of the disorder. When rats receive haloperidol in their drinking water over a prolonged time, at a dose of 1–1.5 mg/kg/day, they develop spontaneous oral chewing movements which appear to be involuntary; these have been called vacuous chewing movements (VCMs) in the literature. Our laboratory has confirmed the gradual onset of these oral movements over the first 6 months of haloperidol treatment [16]. We have confirmed their resistance to anticholinergic administration and to acute neuroleptic withdrawal. Moreover, they are suppressible by acute neuroleptic and by benzodiazepine treatment. Interestingly, the incidence of VCMs varies among rat strains, suggesting a genetic contribution to the drug-induced disorder [16].

Using this animal model, we have demonstrated that the concurrent administration of progabide, a direct-acting $GABA_A$ and $_B$ agonist, prevents the onset of VCMs in rats [17]. Progabide reduces not only the incidence of VCMs, but also their severity. This outcome is consistent with the idea that tardive dyskinesia (or VCMs in this model) is associated with increased numbers of GABA receptors in the SNR, a neuroleptic-induced effect which is antagonized by concurrent progabide administration. But this interpretation was, at the beginning of our most recent experiment, untested and speculative. Other possibilities exist because GABA receptors exist in high density in all the important nuclei of the basal ganglia and thalamus. To answer these questions, we have begun to examine neurochemical correlates of the spontaneous movements in the basal ganglia, including receptor binding for D_1-, D_2- and $GABA_A$-receptors. These studies were carried out in rats treated with haloperidol at a dose of 1.5 mg/kg/day for 6 months. The receptor binding studies have been completed and analyzed. Their results, which are summarized here, suggest a specific hypothesis for further testing.

Twelve rats were treated with haloperidol in drinking water (1.5 mg/kg/day) for 6 months and 7 rats were treated with tap drinking water only; VCMs ratings at the end of the 6-month treatment period were 0.9 ± 0.3 (tap water) and 10.3 ± 3.0 (haloperidol), VCM per 5 min, respectively. Importantly, the haloperidol group was made up of rats with and without these oral movements; 5 rats manifested VCMs (19.5 ± 4.2 per 5 min), while 7 rats did not manifest the VCMs (3.6 ± 0.9 per 5 min). This spontaneous division allows comparisons to be made between haloperidol-treated rats with and without VCMs (i.e. a comparison across behaviors), as well as between haloperidol and water treatment (i.e., a comparison across drug treatment). Methodology for the binding assays is detailed elsewhere [18]; ligands used for the D_1-, D_2- and $GABA_A$-receptors were ^{3}H-SCH23390, ^{3}H-spiperone, and ^{3}H-musimol, respectively.

Data from this experiment has shown that D_2-receptors are increased in striatum in haloperidol-treated rats compared to water alone rats. But, the D_2 upregulation is equal in those rats with and without VCMs (table 2). Thus, the mere upregulation of D_2-receptors in stratum cannot be related to the occurrence of VCMs. In the SNR, $GABA_A$-receptors were increased in haloperidol-treated rats compared to water alone animals. Here, however, the increase was accounted for predominantly, but not exclusively, by the haloperidol-treated rats with VCMs (table 2); rats without VCMs had levels of $GABA_A$-receptors which did not differ significantly from the water only animals. While D_1-receptors in multiple brain regions were unaltered by chronic haloperidol, in the SNR alone, D_1-receptors were reduced by haloperidol, in the animals who displayed VCMs. Haloperidol-treated animals with VCMs had a density of D_1-receptors approximately

Table 2. Effects of haloperidol (6 months) on receptor densities (fmol/mg tissue; mean ± SEM)

	Control (6)	Haloperidol	
		VCMs (5)	no VCMs (7)
D_2 in striatum	72.2±2.0	91.7±4.9*	96.2±3.1**
$GABA_A$ in SNR	302.0±15.9	367.2±7.4*	346±17.5
D_1 in SNR	174.4±9.7	143.5±7.7*	172.9±9.3

* $p < 0.05$; ** $p < 0.01$.

20% lower than the haloperidol-nonVCM rats and water only animals (table 2). The ratio of D_1- to $GABA_A$-receptor number was singularly increased with VCMs, while it was unaffected by drug treatment status [data not shown]. Because the SNR D_1-receptor is located presynaptically to the $GABA_A$-receptor, it would be reasonable to propose that these two receptors would function jointly to regulate GABA release and action in SNR.

These results are consistent with the idea that the resultant synaptic activity influencing $GABA_A$-receptors on GABA-containing nigral long-tract efferent neurons is increased. This activity 'overinhibits' the efferent neurons, reducing inhibition in the nigral target areas. We propose that the combination of $GABA_A$ upregulation and D_1 down-regulation in SNR combines to effectively increase GABA-mediated neurotransmission. Thus, the determinant pathophysiology of haloperidol-induced rat oral dyskinesias resides in the SNR at the GABA synapse on the efferent GABAergic neurons. The mechanism includes upregulated postsynaptic GABA receptors, postsynaptic to the GABA-containing striatonigral pathway, with down-regulated presynaptic D_1-receptors at that same synapse (presumably secondary to dendritic DA release from SN compacta DA containing neurons), resulting in increased GABA-mediated neurotransmission. There is another possibility, that being a degeneration of striatonigral neurons, resulting in $GABA_A$ upregulation (postsynaptic) and D_1 disappearance (presynaptic). This latter possibility is less likely, however, because the VCMs (and TD itself), is slowly reversible over many weeks (months) after neuroleptic withdrawal.

The clinical implications of these findings, once the cause of the D_1-receptor change is determined, could mean the identification of a drug prophylaxis for the disorder. In any case, the identification of the pathophysiology for VCMs, a model for TD, would suggest a screening technique to evaluate which drugs would fail to produce this constellation of neuroleptic effects.

References

1 Delay J, Deniker P, Harl JM: Traitement des états d'excitation et d''agitation par une méthode médicamenteuse dérivée de l'hibernothérapie. Ann Méd Psychol 1952; 10:267–273.

2 Carlsson A, Lindqvist M: Effect of chlorpromazine or haloperidol on formation of 3-methoxytyramine and normetanephrine in mouse brain. Acta Pharmacol Toxicol 1963;20:140–144.

3 Creese I, Burt DR, Snyder SH: Dopamine receptor binding predicts clinical and pharmacological potencies of antischizophrenic drugs. Science 1976;192:481–483.

4 Kane J, Honigfeld G, Singer J, Meltzer H, and the Clozaril Collaborative Study Group: Clozapine for the treatment-resistant schizophrenic: A double-blind comparison with chlorpromazine/benztropine. Arch Gen Psychiatry 1988;45:789–796.

5 Safferman A, Lieberman JA, Kane JM, Szymanski S, Kinon B: Update on the clinical efficacy and side effects of clozapine. Schizophr Bull 1991;17:247–261.

6 Casey DE: Clozapine: Neuroleptic-induced EPS and tardive dyskinesia. Psychopharmacology 1989;99:S47–S53.

7 Farde L, Wiesel FA, Nordstrom AL, Sedvall G: D_1 and D_2-dopamine receptor occupancy during treatment with conventional and atypical neuroleptics. Psychopharmacology 1989;99:S28-S31.

8 Seeman P: Atypical neuroleptics: Role of multiple receptors, endogenous dopamine, and receptor linkage. Acta Psychiatr Scand 1990;82(suppl 358):14–20.

9 Kane JM, Woerner M, Borenstein M, Wegner J, Lieberman J: Integrating incidence and prevalence of tardive dyskinesia. Psychopharmacology 1986;22:254–258.

10 Rupniak NM, Mann S, Hall MD, Fleminger S, Kilpatrick G, Jenner P, Marsden CD: Differential effects of continuous administration for one year of haloperidol or sulpiride on striatal dopamine function in the rat, Psychopharmacology 1984;84: 503.

11 Tarsy D, Baldessarini RJ: Behavioral supersensitivity to apomorphine following chronic treatment with drugs which interfere with the synaptic function of catecholamines. Neuropharmacology 1974;13:927.

12 Thaker GK, Tamminga CA, Alphs LD, Lafferman J, Ferraro TN, Hare TA: Brain aminobutyric acid abnormality in tardive dyskinesia: Reduction in cerebrospinal fluid GABA levels and therapeutic response to GABA agonist treatment. Arch Gen Psychiatry 1987;44:522–529.

13 Gale K: Chronic blockade of dopamine receptors by antischizophrenic drugs enhances GABA binding in substantia nigra. Nature 1980;283:659.

14 Gunne LML, Häggström J-E, Sjöquist B: Association with persistent neuroleptic-induced dyskinesia of regional changes in brain GABA synthesis. Nature 1984;309: 347.
15 Thaker GK, Nguyen JA, Tamminga CA: Saccadic distractibility in schizophrenic patients with tardive dyskinesia. Arch Gen Psychiatry 1989;46:755-756.
16 Tamminga CA, Dale JM, Goodman L, Kaneda H, Kaneda N: Neuroleptic-induced vacuous chewing movements as an animal model of tardive dyskinesia: a study in three rat strains. 1990;102:474.
17 Kaneda K, Shirakawa O, Dale J, Goodman L, Bachus SE, Tamminga CA: Co-administration of progabide inhibits haloperidol-induced oral dyskinesias in rats. Eur J Pharmacol 1992;212:43–49.
18 Shirakawa O, Tamminga CA: Correlation between chronic GABA agonist actions on haloperidol-induced chewing movements and changes in dopamine and GABA-A receptor densities, submitted.

Carol A. Tamminga, MD,
University of Maryland, Maryland Psychiatric Research Center,
PO Box 21247, Baltimore, MD 21228 (USA)

Brunello N, Mendlewicz J, Racagni G (eds): New Generation of Antipsychotic Drugs: Novel Mechanisms of Action. Int Acad Biomed Drug Res. Basel, Karger, 1993, vol 4, pp 77–90

Pharmacological and Psychosocial Long-Term Treatment of Schizophrenia

Steven Hirsch

Department of Psychiatry, Charing Cross & Westminster Medical School, London, UK

After recovery from an acute episode the mainstay of treatment for patients with schizophrenia is continuous maintenance medication with a neuroleptic. Patients will vary in their individual dosage requirements but the variation in medical practice between countries is probably as great as the variation between patients. In the UK more than 70% of patients can be maintained on a depot neuroleptic equivalent to 12.5–25 mg of fluphenazine decanoate every 2 weeks but on double the dose every 4 weeks, while in the USA the usual dose is twice that, so interpretation of research which is dose-dependent can be difficult. Thus, in a number of clinical trials the dose range regarded as a lower dose in the US has overlapped with the normal dose range of studies in the UK. Research involving a comparison of one drug with another is even more problematic because of the difficulty in establishing comparable doses. This is because neuroleptics vary widely in their effective anti-psychotic doses but also differ in their potency on neurotransmitter systems which have other clinical effects – anti-parkinsonism, sedation, etc. Thus two similar drugs like fluphenazine and pimozide are both potent neuroleptics but the anti-cholinergic potency of pimozide is much greater – hence the finding by Falloon et al. [1] of a difference in the effect of pimozide and fluphenazine decanoate on social functioning which can be accounted for by the higher rate of parkinsonian symptoms in the fluphenazine-treated patients.

Davis and Andriukaitis [2] carried out a meta-analysis of 35 controlled studies involving a double-blind comparison of active medication with placebo for long-term treatment of over 3,600 patients in all. The studies varied in the duration of the follow-up from 3 months to 2 years, but the overall relapse rate was 16% for patients continued on neuroleptics

Table 1. Relapse rate following drug discontinuation after long-term remission

	Time in remission, years	Study months	Relapse rate, %
Wistedt, 1981	>6 months	12	100
Hirsch et al., 1973	<1	9	92
Johnson et al., 1976	1–2	6	92
Dencker et al., 1980	2	24	99
Johnson, 1981	1–4	18	80
Hogarty et al., 1976	2–3	12	65
Hirsch et al., 1973	1–5	9	66
Cheung et al., 1980	3–5	18	62

compared to a 57% relapse rate for those treated with placebo. However, even in patients on oral medication relapse was not immediate despite the fact that neuroleptics have been shown on PET scans to disappear from dopamine binding sites within 3 days of being discontinued [Baron, personal commun.]. The mode time for relapse occurs at 3½–4½ months after discontinuing medication [3, 4]. This should provide problems for those who wish to explain the action of neuroleptics simply on the basis of receptor blockade. The greatest risk for relapse is in the first year of treatment after recovery from an acute episode; it then declines exponentially over 2 years, but the risk of relapsing on placebo remains 2½–4 times higher than if patients are maintained on medication [3].

Withdrawal from medication, even when the patients are well selected and have remained well for many years, leads to relapse in most cases. Table 1 summarises several studies in which patients have been withdrawn from medication after remaining well on neuroleptics for between 6 months and 5 years, thus relapse rates may vary from 100 to 63% respectively as this period of previous remission increases. Some authors have suggested that there may be a relationship between higher dosage of treatment and greater risk of relapse [5, 6], but such relationships could be spurious in the sense that a more severe illness with a greater tendency to relapse may also require a higher dosage to control symptoms rather than higher dosage causing the tendency to relapse. Most patients eventually relapse. Extrapolating from their 3-year observation of 374 patients treated on oral medication. Hogarty and Ulrich [3] predicted that 85% of patients on medication and 87% on placebo would eventually relapse. Curson et al.

[7] following up patients from the original placebo-controlled withdrawal study, confirmed this with their 7- to 8-year follow-up of patients on depot fluphenazine. 83% of those well-maintained on neuroleptics have one or more relapses during the period observed; not counting those who died. This compares to the 68% of those treated on placebo who had relapsed by the end of the 1st year. The benefit of maintenance neuroleptic treatment can therefore be best understood as postponing rather than preventing relapse in patients with recurrent or chronic schizophrenia.

Differences between Different Treatments

Despite the differences between the pharmacological profile of neuroleptics, arising from the differences in their potency profile on various neurotransmitters other than the dopamine receptor, there are no replacable findings of a difference between one neuroleptic and another on the core symptoms of schizophrenia prior to the reintroduction of clozapine. This is the one drug which has been proven to be superior to all other neuroleptics in its effect on refractory and negative as well as chronic positive schizophrenic symptoms, an observation that is now well-established [8]. Clozapine is also the one neuroleptic which appears to have little or no tendency to cause tardive dyskinesia and has much less tendency to cause extra-pyramidal symptoms. The down side, as you know, is that it carries a 2–4% risk of causing leucopenia and agranulocytosis which can only be prevented by weekly blood counts. The race is now on to establish an effective clozapine analogue or alternative which does not carry these haematological risks.

Mode of Treatment

From a pharmacological point of view one should not expect depot medication to be more effective than oral medication although it could theoretically overcome bio-availability problems. Closely controlled double-blind clinical studies fail to find any difference in relapse prevention or control of psychotic symptoms between oral and depot maintenance treatments [1, 9, 10]. The study of Schooler et al. [9] included 214 patients randomised to oral or depot fluphenazine and followed them over 1 year; 33% relapsed on oral medication and 24% on depot, a non-significant difference. However, in practice, relapse rates are consistently lower on depot medication [11, 12]. Depot medication gives control over compliance, ensures regular surveillance and immediate feedback if the patients miss their medication, and avoids inadvertent or intended overdosage, and can avert bio-availability problems if they exist. It follows from this discussion that manufacturers of new anti-psychotics should aim to develop potent drugs with a reasonably long half-life which can be incorporated into a depot preparation.

Attempts to Modify the Treatment Approach

From the point of view of short-term toxicity, neuroleptics are among the safest of major drugs in the pharmacopoeia, with very high therapy to toxic effect ratios. Yet we also know that the incidence of less severe side effects in the short term is very high; we found that more than 75% of patients will experience extra-pyramidal side-effects in the first year of maintenance treatment [13]. There is an increasing awareness of the importance of cardiovascular events such as arrhythmia and sudden death, particularly on very high doses of neuroleptics. However, the most important risk is tardive dyskinesia which is thought to develop in more than 20% of patients as they grow older. These concerns have led to a search to modify the concept of maintenance treatment by testing the efficacy of low-dose treatment, one-half to one-tenth the standard dose; and by treating patients only when they first show signs of relapse (Targeted Brief Intermittent Treatment).

Low-Dose Studies

Controlled experiments which attempted to reduce the dose of maintenance medication began as early as 1964 when Caffey et al. [14] compared placebo to daily treatment and treatment 4 times a week, a 43% reduction; Prien et al. [15] used a similar strategy. The risk of relapse was 3–8 times higher with the dose reduction. Capstick [16] reduced dosage by gradually lengthening the time between doses, increasing the interval between visits by 1 week on each visit to a minimum dose of 3 mg per month and stopping after 6 months. 40% of patients relapsed during the dose reduction phase and 40% relapsed over the next 18 months. Notwith-

Table 2. Summary: Low-dose studies

	Dose reduction %	Study months	Relapse rates by treatment, %		Placebo %
			standard	reduced	
Caffey et al., 1964	43	4	5	16	45
Prien et al., 1973	43	4	1	8	
Kane et al., 1983	90	12	7	56	
Marder et al., 1987	80	24	36	69	
Johnson et al., 1987	50	12	10	32	
Hogarty et al., 1988	70	12	14	22	
	48	24	24	30	

standing, interest in dose reduction strategies was raised by Kane et al.'s [8] study of 126 patients comparing a 90% dose reduction to their standard medication which ranged from 25–100 mg per month as compared to UK trials which tend to use a standard dose of 25–50 mg per month. Although low-dose treatment was disappointing in respect to relapse, they did note an improvement in social functioning and fewer signs of tardive dyskinesia. However it could be argued that their improved social functioning was due to a decrease in parkinsonism because the currently used standard dose was too high, and the low dose overcame this. Marder et al. [17], Hogarty et al. [18] and Johnson et al. [19] have all attempted dose reductions of 50–80% from the patients pre-existing maintenance dose and found significant increases in the percent of patients experiencing an exacerbation of their psychotic symptoms. The study of Johnson et al. extended our knowledge by only including patients with a good prognosis and with good social functioning and low levels of symptomatology. Patients on maintenance flupenthixol of not more than 40 mg every 2 weeks (equivalent to 50 mg of fluphenazine decanoate per month) were compared to a half-dose group for 1 year, then patients in the control groups had their original dose reduced by half for a further 2 years. By the end of the trial, relapse had occurred in 56% of those who received half dose for 2 years and 70% of those who were on half dose for 3 years. These studies are scrutinized in table 2.

One may conclude therefore that neuroleptic treatment offers a time- and dose-dependent suppression of certain symptoms of schizophrenic illness. Even a 50% reduction of medication leads to unacceptably high relapse rates over time and side-effects are only mildly improved. However as a short-term strategy, for example during pregnancy, the 30–80% chance of surviving off medication for the first year may be justified.

Brief Targeted Intermittent Treatment

The second approach to reduce exposure to medication derived from Herz and Melville's [20] finding in two surveys that either the patient or a close relative could report notable changes in the patient preceding relapse in 90% of cases [20]. However this was a retrospective study and Wirshing et al. [21] were unable to detect prodromal symptoms among 81 patients when attempting to anticipate relapse, and Jolley et al. [22] could only detect prodromal symptoms before the relapse occurred in about half of their sample in the 2nd year. The hypothesis was that if a psychotic relapse is preceded by an emergence of neurotic, depressive or other symptoms, it was hoped that treatment might be limited to a brief targeted intervention which would be instituted when such prodromal symptoms occurred.

Four trials have now tested this hypothesis. Two involved a placebo substitution for the maintenance medication in a double-blind randomised design [22, 23] and two openly randomised patients to continue previous treatment or only receive treatment for a few weeks when prodromal symptoms occurred [24, 25]. Both Herz et al. [23] and Carpenter et al. [24] gave all patients regular adjunct psychotherapeutic treatment at a level which would be unusual in England. Pietzcker et al. [25] included a third treatment group who only received medication if they had a recurrence of their psychosis. All studies achieved a significant overall reduction in medication for the brief intermittent treatment groups.

The results are convincing; the well-established practice of continuous medication is consistently superior to the targeted brief treatment in all studies. The dropout rates were significantly higher in the intermittent group in all 4 studies. In the study of Herz et al. [23], 46% of patients having brief targeted (intermittent) treatment dropped out, and this seriously distorted the rest of their results as the dropouts were not able to contribute to the relapse ratings and symptom ratings after they were withdrawn. Both Herz et al.'s [23] and our own study [22] recorded the frequency that prodromal symptoms recurred and found a significant increase of prodromal symptoms in patients in the targeted treatment group.

Psychotic symptoms recurred more frequently in both the 1st and 2nd year of Jolley et al.'s [22] study and Carpenter et al.'s [24] study and the 2nd year, if not the 1st, of Pietzcker et al.'s [25], from which we have only had a preliminary report. There are parallel findings in respect to rehospitalisation although the advantage to the continuously medicated group was not significant until the second year in Jolley's study, probably because of the long lasting effect of depot medication used in that study. Effects on social functioning were somewhat inconsistent. Herz noted a significant increase in burden on others in the targeted treatment group and Carpenter noted deterioration in employment status while the other two studies were unable to detect any significant difference.

The conclusion from these studies is that brief targeted treatment carries a significantly higher risk of relapse than continuous maintenance medication and no added risk of side-effects or disadvantage for social functioning. Nevertheless like the part-empty glass, it must be noted that between 37 and 64% of patients survived on the intermittent regime without a relapse.

Low-Dose Top-Up Therapy

There are two studies which suggest what could prove to be an effective and reasonable way to reduce exposure to medication. Marder et al. [17] and Hogarty et al. [18] employed a 50–80% reduction in medication, but allowed clinicians to increase medication at the appearance of minor episodes or the first signs of an emergency relapse. In both studies the recurrence of emergent symptoms was higher in the low-dose group than the occurrence of relapse. In the study of Marder et al. [17] there was a significant difference between low and standard dose *exacerbation* rates, but not between the *relapse rate* at high and low doses. However, it has not been proven that the 'exacerbations' in the boosted up regime were real and would have gone on to relapse had the dose not been increased.

Social Determinants of Relapse

Research suggests that a considerable amount of the variation in the time to relapse after stopping medication can be explained by the effect of social experiences, and that social factors are potent stimulators of relapse, particularly in patients who are not protected by medication. The contribution of social factors on the one hand and prophylactic neuroleptic medication on the other probably varies with the nature of the patient population. Estimates vary, but Davis [26] carried out a meta-analysis of placebo-controlled study and found that the product-moment correlation between continuation on medication and relapse was 0.60, while Vaughan and Leff's [33] estimate from analysing two naturalistic studies of the effect of medication and of family atmosphere characterised as high or low in Expressed Emotion, found the correlation of outcome with medication to be $p = 0.39$ [27]. By these estimates medication explained 36–15% of the variance, respectively; while returning to live with a relative who was rated high or low in Expressed Emotion had a similar correlation, $r = 0.45$, explaining 20% of the variance. These figures must be regarded as estimates only, based on limited observation and a crude analysis which had a high sampling error. Continuation on medication did not depend on the kind of relative, so compliance did not explain the difference betwen the two groups. Thus with some settings at least, the social environment can be an important determinant of relapse. Thus there are various studies which have demonstrated that certain types of social influences put patients at risk for relapse. The two most researched measures are called Expressed Emotion and Life Events.

Expressed Emotion

Expressed Emotion is a measure invented by Rutter and Brown [28] to measure the emotional atmosphere in the home. Bebbington and Kuipers [29] have reviewed 22 studies including further follow-up studies of the same patients, which were carried out in a wide variety of national, racial and cultural settings. Only 3 failed to support the original observation that patients diagnosed as schizophrenic during an acute admission and discharged to live with a spouse or relative rated high in Expressed Emotion had a significantly higher relapse rate in the next 9–12 months – Parker and Hadzi-Pavlovic's [30] analysis of 12 studies involving over 908 patients estimated that the relapse rate of patients returning to live in high Expressed Emotion homes is 3.7 times greater than those going to low Expressed Emotion homes. Expressed Emotion is measured blindly from audio-tapes recorded during an intimate and searching interview [28]. The original studies in the interview took place in the first 2 weeks after the relative's admission to hospital. Salient measures turned out to be the number of hostile critical comments in the first half hour, and evidence of emotional over-involvement such as reluctance to ever allow the adult relative to be alone.

Of the three studies which have failed to find a significantly higher relapse rate in patients who live with High Expressed Emotion relatives, one was based on a survey of patients living in the community – low relapse rates were observed in the High and Low Expressed Emotion groups [31]. Another study by Johnstone and Crow's [32] 'group' had equivocal results but also involved first episode patients, which is different from all the other studies in this field.

The critical experiment to test the validity of High Expressed Emotion as a predictor of early relapse after discharge is a randomised intervention study which controls for medication and clinical variables. Vaughan and Leff [33] were the first of several to publish. All the patients lived in High Expressed Emotion homes. In 8 of 11 prospective intervention studies there was a significantly low relapse rate (about ¼–½) for patients whose High Expressed Emotion relatives had family, behavioural and/or group therapy or some form of psycho-education, although it was often the amount of contact with the relative rather than the level of Expressed Emotion which changed.

The results from prospective controlled intervention studies plus the impressive number of naturalistic or predictive studies strongly support the hypothesis that patients who return to live with High Expressed Emotion with relatives after recovering from a recent episode of schizophrenia will have an increased risk of relapse in the next 2 years. However, this finding is less important in respect to broad implications for causality schizophrenia, than for its heuristic value in showing that schizophrenia is an illness which is susceptible to the influences of psychosocial stress. The relevance of this finding very much varies according to the characteristics of the group studied. In the UK studies, less than two-thirds of schizophrenics lived with relatives, of these about 40% lived with High Expressed Emotion relatives, thus 25% in all; and of these, say 60% will relapse within 9–12 months. That means that 15% have their relapse influenced by Expressed Emotion. Nevertheless, for the individual who is recurrently relapsing and lives with a High Expressed Emotion relative, family intervention treatment may be effective and perhaps life saving.

Life Events

If schizophrenia is a condition which is reactive to psychosocial experience, we would expect that Life Events would impinge on the course of the illness either by precipitating relapse soon after a life event or by a cumulation effect such that the more events one experiences in a period of time, the greater the risk of relapse. This hypothesis has been tested by looking for an increased incidence of schizophrenia following an event, such as induction into the army [34] but most studies have followed Brown and Birley's [35] methodology, which compared the rate of Live Events and the proportion of individuals who experienced a life event in a time period immediately preceding relapse to the rate and number who relapsed in previous time periods or in controls. Despite common beliefs, the evidence supporting Brown's original finding of an increase in Life Events in the 3 weeks just before a relapse is lacking. Of the 12 studies we could find testing the hypothesis that there is a relationship between life events and schizophrenic relapse, only two support the data from Brown and Birley's study which would suggest that Life Events act by provoking or triggering a relapse, while 9 other studies fail to find a higher rate of life events in the immediate pre-relapse period [36].

The evidence that Events have a cumulative effect is stronger. In most of this research Life Events have been defined by a rigorous glossary for rating the likelihood that the event occurred independent of the patient's control and quantified the implied threat, using the Life Event and Difficulty Scale [35] – moving house, redundancy, witnessing a serious accident, and having relations come to stay are examples of the Life Events. However there are conflicting results in previous studies of the rates of Life Events during longer run-up periods to relapse and they are flawed by their retrospective design.

We endeavoured to overcome the problems of most previous studies by carrying out a prospective design including a control for maintenance medication or reduction. There was no significant relationship between the rate of life events in the previous 4 weeks and the likelihood of relapse, but there was a significant relationship between an individual overall relapse rate and the likelihood of relapse. However the effect of medication versus no medication was far stronger and there was no interaction between Life Events and medication. Thus an increase in the rate of life events increased the risk of relapse as much in patients on medication as off. This invalidates the hypothesis we put forward many years ago which suggests that medication raises the threshold to Life Events and social experience [37, 38].

Like the effect of Expressed Emotion, this body of research is more important for its heuristic value – demonstrating that schizophrenia is a condition reactive to social experience – than its value in explaining causation or accounting for the variance in outcome. The impact of Life Events on the cohort studied can be usefully examined in terms of the relative risk of relapsing for those with the average rate of Life Events in the sample, compared to the risks with no Life Events, taken as a value of 1. The relative risk for relapsing with an average number of Live Events in our study was 1.24, and the attributable risk, which represents the added risk for individuals on medication if they had the average rate of Life Events was 19.3%, and 35% if they had twice the average. If all the patients in our sample having Live Events had not had any Life Events, the reduction in the relapse rate for the entire sample would have been only 6.7%.

My conclusion from this work is that the course of schizophrenic illness can be affected by the patients' daily living, social environment and their episodic experience of Live Events and that for some patients these factors can make a big difference, but they do not explain much about the course of the illness for this group as a whole. The effect of medication is much stronger.

Intervention with Social Treatments

The benefit of various forms of family intervention, social support, individual psychotherapy, group therapy, behavioural psychotherapy and social skills have all been reviewed elsewhere. The subject is too detailed and complex to consider here, but certain tentative generalisations may be ventured. A variety of treatments can be shown to be effective in reducing the relapse rate of the group receiving the specific therapy as compared to controls. In several studies the effect of medication and compliance have been controlled and the benefit of the psycho-social approach is significant. The size of the group studied has generally been small and the resulting differences between groups of necessity large. These studies are, in fact, among the best demonstrating that psychotherapeutic and family therapy approaches have an effect.

The down side is that among the 5 controlled studies which have a 2-year follow-up, the effect of treatment over non-treatment has seriously if not wholly diminished, and the add-on effects of social treatments when experimentally combined with ineducation as a variable is not very large. In the 2-year follow-up study of Leff et al. [39] and the recent paper by Hogarty et al. [40] a family psycho-educational approach reduced the relapse rate for patients on maintenance medication living in High Expessed Emotion families by half to patients receiving psycho-education in comparison. The addition of Social Skills training reduced it further by half so that the treated group had no relapses at the end of year one; however by the end of the second year the benefit of social skills training was lost. Elsewhere Leff's group have shown that a group educational approach for relatives was as effective as treatment of individual families for those who would attend [38]. These studies show that in combination with medication, pschosocial treatments can have a time-limited effect on the risk of relapse and imply that, where possible, treatments needs to be renewed, though the benefit of such a regime has never been tested. For those societies which can afford it and those families that will cooperate, they offer a worthwhile advantage. Given the much more powerful effect of medication and the advances in chemotherapeutics we look forward to in the future, hopefully these more expensive and timeconsuming adjuncts to treatment will become less and less necessary.

References

1 Falloon I, Watt DC, Shepherd M: The social outcome of patients in a trial of long-term continuation therapy in schizophrenia and pimozide vs fluphenazine IM. Psychol Med 1978;8:265–274.

2 Davis JM, Andriukaitis S: The natural course of schizophrenia and effective maintenance treatment. J Chem Psychopharmacol 1988;6:25.

3 Hogarty GE, Ulrich R: Temporal effects of drug and placebo in delaying relapse in schizophrenic outpatients. Arch Gen Psychiatry 1977;34:297–301.

4 Hirsch SR, Gaind R, Rohde PD, Stevens BC, Wing JC: Outpatient maintenance of chronic schizophrenic patients with long-acting fluphenazine: Double-blind placebo trial. Br Med J 1973;i:633–637.

5 Prien RF, Cole JO: High dose chlorpromazine therapy in chronic schizophrenics. Arch Gen Psychiatry 1968;18:482–495.

6 Wistedt B: A depot neuroleptic withdrawal study: A controlled study of the clinical effects of the withdrawal of depot fluphenazine decanoate and depot flupenthixol decanoate in chronic schizophrenic patients. Acta Psychiatr Scand 1981;64:65–84.

7 Curson DA, Barnes TRE, Bamber R, Platt SD, Hirsch DRH, Duffy JA: 7-Year follow-up study of MRC 'Modecate' trial. Br J Psychiatry 1985;146:464–480.

8 Kane J, Rifkin A, Woerner M, Reardon G, Sarrantakos S, Schiebel D, Ramos Lorensi J: Low dose neuroleptic treatment of out-patient schizophrenics. Arch Gen Psychiatry 1983;40:893–896.

9 Schooler NR, Levine J, Severe JB, Brauzer B, DiMascio A, Klerman GL, Tuason VB: Prevention of relapse in schizophrenia: an evaluation of fluphenazine decanoate. Arch Gen Psychiatry 1980;37:16–24.

10 Barnes TRE, Milavic G, Curson DA, Platt SD: Use of the Social Behaviour Assessment Schedule (SBAS) in a trial of maintenance antipsychotic therapy in schizophrenic outpatients: Pimozide versus fluphernazine. Soc Psychiatry 1983;18:193–199.

11 Babiker IE: Comparative efficacy of long-acting depot and oral neuroleptic medications in preventing schizophrenic recidivism. J Clin Psychiatry 1987;48:94–97.

12 Johnson DAW: Further observations on the duration of depot neuroleptic maintenance therapy in schizophrenia. Br J Psychiatry 1979;135:524–530.

13 Knights A, Okasha MS, Salih M, Hirsch S: Depressive and extra-pyramidal symptoms and clinical effects: A trial of fluphenazine versus flupenthixol. Br J Psychiatry 1979;135:515–524.

14 Caffey EM, Diamond LS, Frank TV, Grasberger JC, Herman L, Klett CJ, Rothstein R: Discontinuation or reduction of chemotherapy in chronic schizophrenics. J Chronic Dis 1964;17:374–459.

15 Prien RF, Gillis RD, Caffey EM: Intermittent pharmacotherapy in chronic schizophrenia. Hosp Community Psychiatry 1973;24:317–322.

16 Capstick N: Long-term fluphenazine decanoate. Maintenance dosage requirements of chronic schizophrenic patients. Acta Psychiatr Scand 1980;61:256–262.

17 Marder SR, Van Putten T, Mintz J, Lebell M, McKenzie J, May PR: Low and conventional-dose maintenance therapy with fluphenazine decanoate: Two-year outcome. Arch Gen Psychiatry 1987;44:518–521.

18 Hogarty GE, McEvoy JP, Munetz M, et al: Dose of fluphenazine, familial expressed emotion and outcome in schizophrenia. Arch Gen Psychiatry 1988;45:797–805.

19 Johnson DAW, Ludlow JM, Street K, Taylor RDW: Double-blind comparison of half-dose and standard-dose flupenthixol decanoate in the maintenance treatment of stabilised outpatients with schizophrenia. Br J Psychiatry 1987;151:634–638.

20 Herz MI, Melville C: Relapse in schizophrenia. Am J Psychiatry 1980;137:801–805.

21 Wirshing WC, Marder SR, Johnstone-Cronk M, Lebell J, Mackenzie J, Mintz J, Eckman T, Liberman RP: Management of risk of relapse in schizophrenia. Schizophr Res 1992;6:107–108.

22 Jolley AG, Hirsch Sr, Morrison E, McRink A, Wilson L: Trial of brief intermittent neuroleptic prophylaxis for selected schizophrenic outpatients: clinical and social outcome at two years. Br Med J 1990;301:837–842.

23 Herz M, Glazer WM, Moster MA, Sheard MH, Szymanski H, Hafez H, Mriza M, Vaha J: Intermittent vs maintenance medication in schizophrenia. Two year results. Arch Gen Psychiatry, in press.

24 Carpenter W, Hanlon T, Heinrichs D, Kirkpatrick B, Levine J, Buchanan R: Continuous versus targeted medication in schizophrenic patients: outcome results. Am J Psychiatry 1990;147:1138–1148.

25 Pietzcker A, Gaebel W, Kopcke W, Linden M, Muller P, Muller-Spohn F, Schussler G, Tegeler J: A German multi-centre study on the neuroleptic long-term therapy of schizophrenic patients: Preliminary report. Pharmacopsychiatry 1986;19:161–166.
26 Davis JM: Overview: Maintenance therapy in psychiatry. I Schizophrenia. Am J Psychiatry 1975;132:1237–1245.
27 Hirsch SR: Influence of social experience and environment on the course of schizophrenia; in Hirsch SR, Bradley PB (eds): The Psychopharmacology and Treatment of Schizophrenia. Oxford, Oxford University Press, 1986, pp 200–212.
28 Rutter M, Brown GW: The reliability and validity of measures of family life and relationships in families containing a psychiatric patient. Soc Psychiatry 1966;1: 38–53.
29 Bebbington P, Kuipers E: Social causation of schizophrenia in Bhugra D, Leff JP (eds): Principles of Social Psychiatry. Oxford, Blackwells, in press.
30 Parker G, Hadzi-Pavlovic D: Expressed Emotion as a predictor of schizophrenic relapse: An analysis of aggregated data. Psychol Med 1990;20:961–965.
31 McCreadie RG, Wils D, Grant S, Crockett GT, Mahmood Z, Livingston MG, Watt JA, Greene JG, Kershaw PW, Todd N: The Scottish First Episode Schizophrenia Study V. One year follow-up. Br J Psychiatry 1988;152:470–476.
32 Johnstone EC, MacMillan JF, Frith CD, Benn DK, Crow TJ: Further investigations of the predictors of outcome following first schizophrenic episodes. Br J Psychiatry 1990;157:182–189.
33 Vaughan C, Leff J: The influence of the family and social factors on the course of psychiatric illness. Br J Psychiatry 1976;129:125–137.
34 Steinberg H, Durrell J: A stressful social situation as a precipitant of schizophrenic symptoms: An epidemiological study. Br J Psychiatry 1968;114:1097–1105.
35 Brown GW, Birley JLT: Crises and life changes and the onset of schizophrenia. J Health Soc Behav 1968;9:203–214.
36 Hirsch SR, Cramer P, Bowen J: The triggering hypothesis of the role of life events in schizophrenia. Br J Psychiatry 1992;161(Suppl 18):84–87.
37 Leff JP, Hirsch S, Gainde R, Rohde P, Steven B: Live events and maintenance therapy in schizophrenic relapse. Br J Psychiatry 1973;123:659–680.
38 Leff JP, Kuipers L, Vaughan C, Sturgeon D: Life events, relatives, expressed emotion and maintenance neuroleptics in schizophrenic relapse. Psychol Med 1983;13: 799–806.
39 Leff JP, Kuipers L, Berkovitz R, Ebstein-Vries R, Sturgeon D: A controlled trial of social intervention in the families of schizophrenic patients. Br J Psychiatry 1982; 141:121–134.
40 Hogarty GE, Anderson CM, Reiss DJ, Kornblith SJ, Greenwald DP, Ulrich RF, Carter M: Family psychoeducation, social skills training, and maintenance chemotherapy in the aftercare treatment of schizophrenia. II Two-year effects of a controlled study on relapse and adjustment. Arch Gen Psychiatry 1991;48:340–347.

Prof. Steven Hirsch, Department of Psychiatry, Charing Cross Hospital & Westminster Medical School, St. Dunstan's Road, GB–London W6 8RP (UK)

Brunello N, Mendlewicz J, Racagni G (eds): New Generation of Antipsychotic Drugs: Novel Mechanisms of Action. Int Acad Biomed Drug Res. Basel, Karger, 1993, vol 4, pp 91–96

Clinical-Biochemical Investigations with Classical and Nonclassical Neuroleptics

M. Ackenheil

Department of Neurochemistry, Psychiatric Hospital of the University of Munich, FRG

The definition of atypical antipsychotics includes improvement of schizophrenic symptoms without the typical, unwanted motor side effects of Parkinson's syndrome, dystonia, and akathisia. Effective treatment of symptoms, mostly less responding to typical antipsychotics, is a further characteristic. Antipsychotic drugs have a variety of actions on different neurotransmitter systems and their receptors. The D_2 dopamine receptor blockade is a characteristic common to all effective antipsychotics used today in clinical practice. There is even a correlation between the potency of D_2-receptor blockade and clinical dosage [1]. Most probably, the pharmacological clinical profile results from the biochemical potency in which other dopamine receptors and other receptors, e.g. noradrenaline, acetylcholine, and serotonin (fig. 1), are influenced, too. Atypical antipsychotics like clozapine or zotepine [2, 3], in comparison to typical antipsychotics such as haloperidol, show a stronger action on noradrenaline and serotonin receptors. However, the most ancient typical antipsychotic chlorpromazin has a similar biochemical profile.

Clozapine which, due to clinical results [4] was first described as an atypical antipsychotic as it showed good antipsychotic activity without EMP side effects [5], is, on the basis of animal experiments as well as of PET investigations [6], characterized by a relatively high potency of D_1 in relation to D_2 dopamine receptor blockade. A similar result was found with zotepine, but not with haloperidol [7]. The blockade of D_4 dopamine receptors which were recently cloned [8] is discussed as being important for the effects of clozapine. But until now it could not be clarified whether the unequal potency of dopamine receptor subtype blockade or the addi-

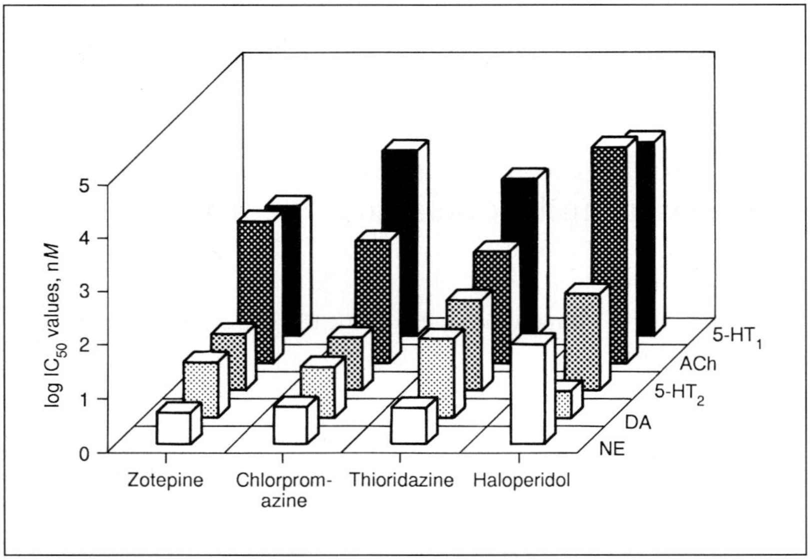

Fig. 1. The biochemical profile of zotepine, haloperidol, thioridazine and chlorpromazine calculated according to results of Harada et al. [23].

tional blockade of α-receptors and serotonin receptors, the preferential action on the mesolimbic dopamine system [9], or other effects may account for the atypical profile of clozapine. Clinical investigations are faced with the problem that patients with the diagnosis of schizophrenia or schizoaffective psychosis are no homogeneous group of patients. They present different psychopathologic patterns and the course of the illness may vary as well. Different factors, e.g. genetic vulnerability, psychosocial environment, and stressful life events may be responsible for the exacerbation of the illness. In clinical practice target symptoms are treated and the therapeutic outcome is dependent on the interaction of the individual patient and the pharmacological treatment which again depends on dosage, pharmacokinetics, and pharmacodynamics of the drug.

Clinical-biochemical investigations with clozapine in comparison with typical antipsychotics like haloperidol have shown that in CSF the dopamine metabolite homovanillic acid (HVA) as well as the serotonin metabolite 5-hydroxyindoleacetic acid are increased and the noradrenaline

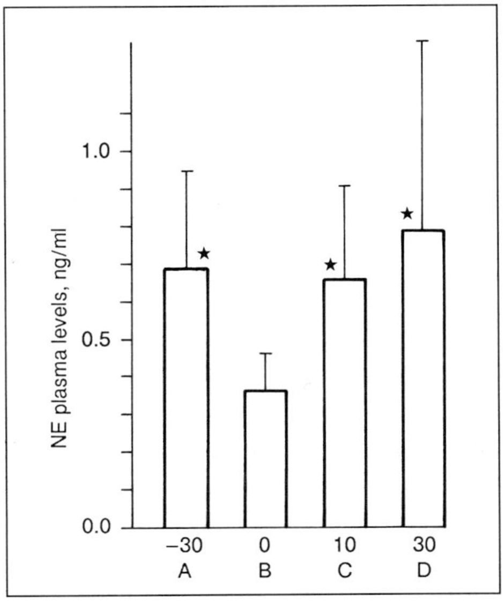

Fig. 2. Plasma noradrenaline (NE) after long-term treatment with typical antipsychotics (A) after 30 days of withdrawal (B) and during 10 (C) and 30 (D) days of treatment with 600 mg sulpiride/day [24, 25].

metabolite 3-methoxy-4-hydroxyphenylglycol is decreased [10]. In plasma a decrease of HVA, a slight increase of prolactin, and a pronounced elevation of noradrenaline were observed [7]. The elevation of noradrenaline indicates an α_1-receptor blockade of clozapine which is stronger than that found with typical antipsychotics. Whereas the dopamine receptor blockade indicated by e.g. CSF HVA shows an adaptation, the noradrenaline levels are still elevated after long-term treatment with neuroleptics [11]. This noradrenaline elevation is normalized after withdrawal of long-term treatment (fig. 2). but increases again after retreatment. Due to these findings it is concluded that the noradrenaline system may be important for the clinical effect as well and may even play a role in the pathophysiology of the schizophrenic illness [12].

The measurement of plasma (p) HVA allows the evaluation of the dopamine metabolism in the limbic system of man [13, 14]. Pickar et al. [15] reported a good correlation of pHVA values and clinical symptomatology in schizophrenic patients. In a similar study carried out in Munich with typical and with novel drugs in development, which were thought to be atypical, again a similar decrease of pHVA after 4 weeks of treatment was found. This decrease was correlated to the improvement measured with the BPRS. Pretreatment pHVA levels in female schizophrenic patients were higher compared to controls. Patients medicated with benzodiazepines during the washout period had lower pHVA levels indicating that this nonspecific treatment may reduce the dopamine turnover. The most interesting finding was the bipolar course of pHVA during the first week of treatment with typical and atypical antipsychotics. Typical antipsychotics, butyrophenones and phenothiazines, induced an increase of pHVA similar to the increase described after a single administration of haloperidol [16]. Later these elevated pHVA levels diminish below pretreatment levels. In a subgroup of patients, pHVA levels decreased continuously. These patients were mostly treated with novel antipsychotics, which are thought to be atypical [17]. Zotepine, the Sandoz compound SDZ HDC 912, savoxepine (Ciba), and amperozide (Leo) belong to this group of drugs. This decrease can be explained by a reduced dopamine release from the presynaptic site. However, it should be mentioned that EPMS side effects occurred in some patients during treatment with the novel drugs [18]. Thus, according to the clinical definition the drugs are not atypical. According to their assumed mechanism of action at the presynaptic site they are atypical.

In a clinical study carried out by Bowers et al. [19], elevated pHVA was reported to be a predictor of rapid, good response to haloperidol treatment. Our results have not conclusively supported this finding so far, for male schizophrenic patients presented more severe symptoms than female patients who in turn had higher pHVA levels. The reason for these discrepancies could be the heterogeneity of the illness. Therefore, it would be well worth subdividing schizophrenic patients according to biological markers. One of these markers could be the elevated spiperone binding on lymphocytes [20]. The variation in binding capacity data was proposed to be predictive of therapeutic response [21]. The retrospective evaluation of several clinical parameters and therapeutic outcome, carried out in 35 schizophrenic patients, supported this publication [22]. Concerning the response to treatment it turned out that schizophrenic patients with low or normal spiperone binding on lymphocytes did not show a good response to treatment with classical antipsychotics. According to this result it is suggested that spiperone binding is not only able to discriminate between biologically different subtypes of schizophrenia, but might also be a predictive variable for response to antipsychotic treatment.

At present, these studies are still in their initial phase. In future, it will be necessary to identify subgroups of schizophrenic patients to better understand the pathophysiology, thus enabling a more specific treatment with typical and atypical antipsychotics.

References

1 Seeman P, Lee T: Antipsychotic drugs: Direct correlation between clinical potency and presynaptic action of dopamine neurones. Science 1975;188:1217–1219.

2 Ackenheil M: Das biochemische Wirkprofil von Zotepin im Vergleich zu anderen Neuroleptika. Fortschr Neurol Psychiatr 1991;1:2–9.

3 Dieterle D, Ackenheil M, Müller-Spahn F, Kapfhammer H-P: Zotepine, a neuroleptic drug with a bipolar therapeutic profile. Pharmacopsychiatry 1987;20:52–58.

4 Angst J, Bente D, Berner P, Heimann H, Helmchen H, Hippius H: Das klinische Wirkungsbild von Clozapin (Untersuchungen mit dem AMP-System). Pharmacopsychiatry 1971;4:202–211.

5 Ackenheil M, Hippius H: Clozapine; in Usdin EB, Forrest I (eds): Psychotherapeutic Drugs. New York, Dekker, 1977, pp 923–956.

6 Farde L, Wiesel FA, Nordström AL, Sedvall G: D1- and D2-dopamine receptor occupancy during treatment with conventional and atypical neuroleptics. Psychopharmacology 1989;99:28-31.

7 Ackenheil M: Clozapine. Pharmacokinetic investigations and biochemical effects in man. Psychopharmacology 1989;99:32–37.

8 Van Tol HHM, Bunzow JR, Guan H-Ch, Sunahara RK, Seeman P, Niznik HB, Civelli O: Cloning of the gene for a human dopamine D4 receptor with high affinity for the antipsychotic clozapine. Nature 1991;350:610–614.

9 Anden NE, Stock G: Effect of clozapine on the turnover of dopamine in the corpus striatum and the limbic system. J Pharm Pharmacol 1973;25:346–348.

10 Ackenheil M, Beckmann H, Greil W, Hoffmann G, Markianos E, Raese J: Antipsychotic efficacy of clozapine in correlation to changes in catecholamine metabolism in man; in Forrest IS, Carr CJ, Usdin E (eds): The Phenothiazines and Structurally Related Drugs. New York, Raven, 1974.

11 Naber D, Finkbeiner B, Fischer B, Zander K, Ackenheil M: Effect of long-term neuroleptic treatment on prolactin and norepinephrine levels in serum of chronic schizophrenics: Relations to psychopathology and extrapyramidal symptoms. Neuropsychobiology 1980;6:181–189.

12 Ackenheil M, Albus M, Müller F, Müller Th, Welter D, Zander K, Engel RR: Catecholamine response to short-time stress in schizophrenic and depressive patients; in Usdin E, Kopin I, Barchas JD (eds): Catecholamines: Basic and Clinical Frontiers. Oxford, Pergamon, 1979, pp 1937–1939.

13 Bowers MB, Heninger GR, Sternberg D, Meltzer HY: Clinical processes and central dopaminergic activity in psychotic disorders. Commun Psychopharmacol 1980;4: 177.

14 Cutler NR, Jeste DV, Karouen F, Wyatt RJ: Low dose apomorphine reduces serum homovanillic acid concentrations in schizophrenic patients. Life Sci 1982;30:753.

15 Pickar D, Labarca R, Linnoila M, Roy A, Hommer D, Everett D, Paul S: Neuroleptic-induced decrease in plasma homovanillic acid and antipsychotic activity in schizophrenic patients. Science 1984;225:954–956.

16 Davila R, Manero E, Zumarraga M, Andia I, Schweitzer JW, Friedhoff AJ: Plasma homovanillic acid as a predictor of response to neuroleptics. Arch Gen Psychiatry 1988;45:564–567.

17 Dengler FX, Kurtz G, Ackenheil M: Plasma-Homovanillinsäure und psychopathologisches Zustandsbild bei schizophrenen Patienten unter Neuroleptikabehandlung (Abstract). DGBP-Kongress, Berlin 1990.

18 Naber D, Gaussares C, Moeglen JM, Tremmel L, Bailey PE: Efficacy and tolerability of SDZ HDZ 912, a partial dopamine D2 agonist, in the treatment of schizophrenia; in Meltzer HY (ed): Novel Antipsychotic Drugs. New York, Raven, 1992, pp 99–107.

19 Bowers M, Swigar ME, Jatlow PI, Goicoechea N: Plasma catecholamine metabolites and early response to haloperidol. J Clin Psychiatry 1984;45:248–251.

20 Bondy B, Dengler FX, Oertel WH, Ackenheil M: 3H-Spiperone binding to lymphocytes is increased in schizophrenic patients and decreased in Parkinson patients; in Przutek H, Riederer P (eds): Early Diagnostic and Preventive Therapy in Parkinson's Disease. Berlin, Springer, 1989, pp 205–212.

21 Grodzicki J, Pardo M, Schved G, Schlosberg A, Fuchs S, Kanety H: Differences in 3H-spiperone binding to peripheral blood lymphocytes from neuroleptic responsive and nonresponsive schizophrenic patients. Biol Psychiatry 1990;27:1327–1330.

22 Bondy B, Ackenheil M, Birzle W, Engel RR, Ertl M: Binding of 3H-spiperone to mononuclear cells of schizophrenics: Relations to clinical variables and outcome (abstract). Eur Neuropsychopharmacol 1991;1:459–460.

23 Harada T, Ebara T, Otsuki S: Possible relationship between antimanic effect and activity of zotepine to 5-HT1 receptor. Folia Psychiatr Neurol Jpn 1984;38:528.

24 Ackenheil M, Fischer B, Hertting N, Zimmer R: Die Wirkung von Sulpirid auf GABA sowie auf den Dopamin- und Noradrenalin-Umsatz beim Menschen. Arzneimittelforschung 1980;30:1205.

25 Zander K, Fischer B, Zimmer R, Ackenheil M: Long-term neuroleptic treatment of chronic schizophrenic patients: Clinical and biochemical effects of withdrawal. Psychopharmacology 1981;73:43–47.

Prof. Dr. M. Ackenheil, Department of Neurochemistry, Psychiatric Hospital of the University of Munich, Nussbaumstrasse 7, D-W-8000 Munich (FRG)

Brunello N, Mendlewicz J, Racagni G (eds): New Generation of Antipsychotic Drugs: Novel Mechanisms of Action. Int Acad Biomed Drug Res. Basel, Karger, 1993, vol 4, pp 97–102

Remoxipride and Haloperidol in Acute Phase Schizophrenia – Experiences from a Recent Clinical Trial Programme

Tommy Lewander, Sven-Erik Westerbergh, Stephan Ogenstad, Bengt Gustafsson

Astra Arcus AB, Södertälje, Sweden

Introduction

The chemical and pharmacological characteristics of remoxipride, (S)-(–)-3-bromo-2,6-dimethoxy-N-[(1-ethyl-2-pyrrolidinyl)-methyl] benzamide hydrochloride monohydrate, have been described previously [1–4]. The clinical documentation of the antipsychotic efficacy and side-effect profile of remoxipride includes nine double-blind comparisons versus haloperidol [5]. Patients from these studies have been pooled and analysed collectively in order to contrast the clinical profile of remoxipride against haloperidol, a well-known classical neuroleptic. This analysis showed that the two compounds were approximately equal regarding antipsychotic efficacy, while remoxipride had a lower frequency and severity of extrapyramidal symptoms (EPS) and less use of concomitant anticholinergics than haloperidol. A number of descriptive post hoc analyses have been made utilizing this patient material. Thus, first episode patients with schizophrenia (approximately 10% of the total patient material) responded favourably to both treatments regarding efficacy. The differences in EPS and the use of concomitant anticholinergics were in favour of remoxipride and more pronounced in this patient group [6]. A third of the total patient material fulfilled criteria for severe psychotic symptomatology [7]. Also in this selected group, there was no difference in efficacy between the 2 drugs

[8]. Other important issues in the treatment of schizophrenia are the response rate, the onset of action and whether or not there is a differential response to different antipsychotics regarding specific symptoms of schizophrenia. In order to shed light on these issues, the same patient material as described in Lewander et al. [5] was utilized.

Materials and Methods

Patients with schizophrenia (DSM III criteria) in an acute phase of the illness (first episode or acute exacerbation of a chronic condition) having received treatment with remoxipride (n = 536; dose range 120–600 mg/day) or haloperidol (n = 396; dose range 5–45 mg/day) under double-blind conditions for at least 12 days up to 4–6 weeks were selected. The patients were divided into four response categories depending on their percentage change of the total Brief Psychiatric Rating Scale (BPRS) [9] scores from baseline to the last available rating. Non-responders were defined as a less than 25% reduction partial responders as a 25–49% reduction, good responders as a 50–74% reduction and very good responders as more than 75% reduction in total BPRS scores, respectively. Overall, 28% of the patients were non-responders, 24% were partial responders, 31% good responders and 17% were very good responders. There were no clinically relevant differences in the distribution of patients to these response categories between to two treatment groups.

Results

As shown in figure 1 there was no difference in the total BPRS score between the response categories, or treatment groups, at baseline. The time courses for improvement (decrease in total BPRS score) in partial, good and very good responders displayed a clear overall improvement, although to different degrees, already within 1–2 weeks, continuing over the next 2–4 weeks. The non-responders did not change appreciably over this time period. There were no clinically relevant differences between the two treatments. In a subsequent analysis, each individual item in the BPRS was averaged for the respective treatment and response category at baseline and at the last rating. As shown in figure 2a, the very good responders showed a reduction of every item in the BPRS to a similar degree for the two drug treatments. The partial and good responders showed a graded but similar reduction of all items with no difference between the drugs [data not shown]. The non-responders did not change in any BPRS item and again there was no difference between the two treatments (fig. 2b).

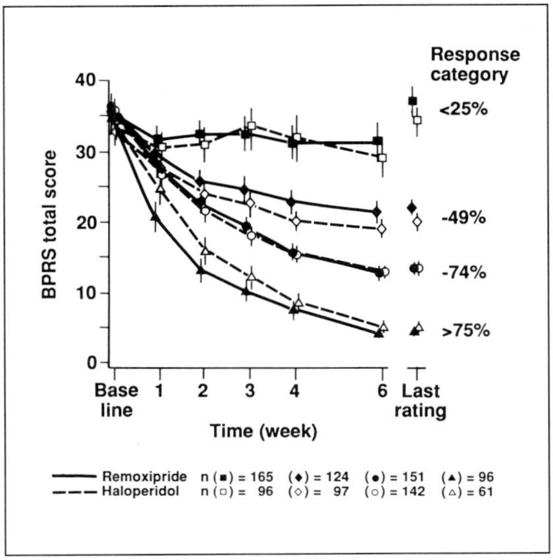

Fig. 1. Effects of remoxipride and haloperidol on BPRS total scores over time in four treatment response categories defined according to the percentage change at the last available rating (last rating divided by the baseline rating) in patients in an acute phase of schizophrenia (> 75% change = very good, 50–74% = good, 25–49% = partial, < 25% = non-responders).

Discussion

The antipsychotic efficacy of remoxipride has been established in a series of clinical studies. The low EPS profile of remoxipride could be consistently verified in several comparative studies versus haloperidol [5]. The focus of the present communication is on aspects of antipsychotic efficacy of the novel low-EPS antipsychotic remoxipride in comparison with the classical neuroleptic haloperidol. Conventional analyses of efficacy in groups of patients in an acute phase of schizophrenia, dominated by positive symptoms, showed apparently similar degrees of improvements for the two compounds [5, 6, 8]. The present detailed analysis of response categories and individual symptoms within the syndrome of schizophrenia has not disclosed any differences between the 2 drugs regarding their efficacy profiles. Thus, these results indicate that a strong propensity for inducing EPS, as for haloperidol, does not seem to carry

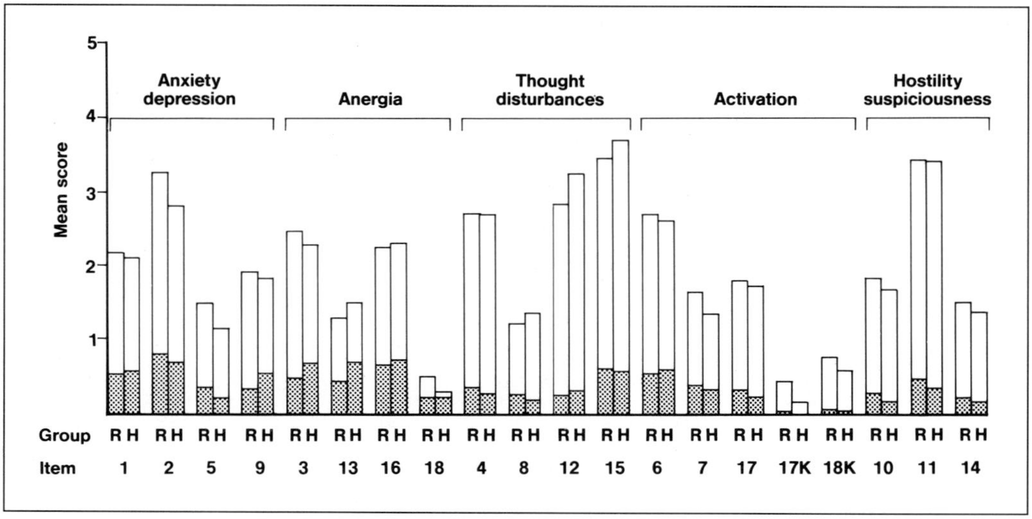

a

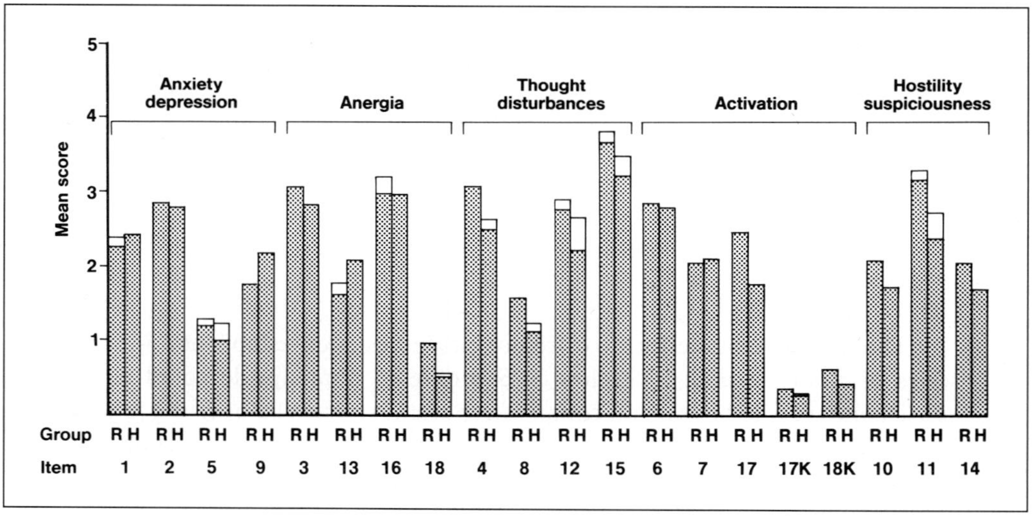

b

Fig. 2. BPRS item analysis in patients being treated with remoxipride (R) and haloperidol (H), respectively. Open columns represent the mean score at baseline; filled columns represent the mean score at the last available rating. *a* Patients displaying a very good response (> 75% change in the total BPRS score, cf. fig. 1). *b* Patients being non-responders (< 25% change). Item numbers refer to the items as defined in the BPRS, and they are arranged according to the BPRS factors described in Overall and Gorham [9]. BPRS-K refers to the BPRS as modified by Krawiecka et al [11], which was used in one of the nine studies [5].

better antipsychotic efficacy in patients with a severity of illness varying from moderate to extreme according to the Clinical Global Impression Scale. Both the new and the classical compound induced improvement of positive, negative and other symptoms.

The present analysis has pointed at some important aspects of drug treatment of acute-phase schizophrenia. First, 25–30% of the patients do not respond to the 2 antidopaminergic drugs within 4–6 weeks of treatment. There are indications that non-responders to haloperidol have similar plasma concentrations and similar degrees of dopamine D_2-receptor occupancy as responders [10]. This group of patients is most probably in need of a different treatment. Secondly, the present data suggest that a response to treatment affects all symptoms of the disorder to a similar degree, at least for the compounds studied. It remains to be investigated, whether or not other antipsychotics with different pharmacological properties also have different efficacy profiles in this type of analysis. Third, the onset of the antipsychotic action in responders seems to be surprisingly rapid, i.e. within 1–2 weeks. This result is in contrast to the general opinion expressed in many textbooks that the onset of action is slow, i.e. several weeks to months. This is certainly not the case with the majority of patients in an acute phase of schizophrenia. These results raise questions on important clinical issues, e.g. when to switch to other compounds if a patient turns out to be a true non-responder. Also, the present analyses have increased our insight in treatment outcome in schizophrenia, which may be of value in the development of new generations of antipsychotic drugs.

References

1 Florvall L, Ögren S-O: Potential neuroleptic agents: 2,6-dialkoxy benzamide derivatives with potent dopamine receptor blocking activities. J Med Chem 1982;25: 1280–1286.

2 Ögren S-O, Hall H, Köhler C: Remoxipride, a new potential antipsychotic compound with selective antidopaminergic actions in the rat brain. Eur J Pharmacol 1983;102:459–474.

3 Ögren S-O, Florvall L, Hall H, Magnusson O, Ängeby-Möller K: Neuropharmacological and behavioural properties of remoxipride in the rat. Acta Psychiatr Scand 1990;82(suppl 358):21–26.

4 Köhler C, Hall H, Magnusson O, Lewander T, Gustafsson K: Biochemical pharmacology of the atypical neuroleptic remoxipride. Acta Psychiatr Scand 1990;82(suppl 358):27–36.

5 Lewander T, Westerbergh S-E, Morrison D: Clinical profile of remoxipride – a combined analysis of a comparative double-blind multicentre trial programme. Acta Psychiatr Scand 1990;82(suppl 358):92–98.
6 Gustafsson B, Blomqvist M, Lewander T, Ogenstad S: Remoxipride in the treatment of first episode of schizophrenia (abstract 27). Psychopharmacology 1991;103:B9.
7 Dubin WR, Waxman HM, Weiss KJ: Rapid tranquilization: The efficacy of oral concentrate. J Clin Psychiatry 1985;46:475–478.
8 Lewander T, Blomqvist M, Gustafsson B: Remoxipride in patients with behavioural dyscontrol (abstract 28). Psychopharmacology 1991;103:B9.
9 Overall JE, Gorham DR: BPRS. Brief Psychiatric Rating Scale; in Guy W (ed): ECDEU assessment manual for psychopharmacology, revised edition. Rockville, National Institute of Mental Health, 1976, pp 157–169.
10 Wolkin A, Barouche F, Wolf AP, Rotrosen J, Fowler JS, Shiue C-Y, Cooper TB, Brodie JD: Dopamine blockade and clinical response: Evidence for two biological subgroups of schizophrenia. Am J Psychiatry 1989;146:905–908.
11 Krawiecka M, Goldberg D, Vaughan M: A standardized psychiatric assessment scale for rating chronic psychotic patients. Acta Psychiatr Scand 1977;55:299–308.

Tommy Lewander, MD, Astra Arcus AB, S–151 85 Södertälje (Sweden)

Brunello N, Mendlewicz J, Racagni G (eds): New Generation of Antipsychotic Drugs: Novel Mechanisms of Action. Int Acad Biomed Drug Res. Basel, Karger, 1993, vol 4, pp 103–117

Utilization, Multidimensional Outcome Measures, and Risk-Benefit Considerations for the Use of Clozapine in Schizophrenia

Herbert Y. Meltzer

Departments of Psychiatry and Pharmacology, Case Western Reserve University, School of Medicine, Cleveland, Ohio, USA

Introduction

Treatment of schizophrenic patients with clozapine is reserved for treatment-resistant and neuroleptic-intolerant schizophrenic patients [1–4] for whom it is usually the treatment of choice. The reason for not using it as a first-line drug is that clozapine produces agranulocytosis in 1–2% of patients. Although the morbidity and mortality from agranulocytosis can be dramatically reduced by weekly monitoring of the total white count or neutrophil count, it is the current practice of most psychiatrists in countries where clozapine is approved not to utilize it to treat patients who can tolerate typical antipsychotic drugs and who have a satisfactory response to them. However, it must be kept in mind that what is a satisfactory response to neuroleptic treatment has a subjective component [5]. The opinions of patients and their families about the satisfactoriness of their response to neuroleptic treatment should be respected in considering whether or not clozapine treatment should be initiated.

The purpose of this article is to consider the optimal use of clozapine in treatment-resistant patients and to present the argument for a multidimensional approach to evaluating the outcome of treatment. The latter issue has been discussed in more detail elsewhere [4].

Treatment Resistance and Neuroleptic Intolerance

Treatment resistance may be defined as failure to respond to adequate clinical trials of 3 typical antipsychotic drugs of at least two different classes [1]. Ordinarily, patients will have been ill for more than 2 years before this situation arises because the large majority of schizophrenic patients will respond to antipsychotic drugs during the first few psychotic episodes. However, all clinicians have treated patients whose initial psychotic symptoms never respond adequately to treatment from the beginning of their illness. These individuals could be candidates for clozapine within the first 6 months of treatment. However, the duration of illness in the typical clozapine patient is likely to be between 5 and 40 years when clozapine treatment is first initiated in a country where it has not previously been approved. Most treatment-resistant cases will have a mixture of positive symptoms, negative symptoms and disorganization (incoherence, loose associations, inappropriate affect, poverty of thought content) [6]. Patients who show only one of these three types of symptoms are still candidates for clozapine treatment. It is definitely useful to consider clozapine treatment for patients who have few positive symptoms but have extensive negative symptoms [7].

Clozapine should be considered in neuroleptic-intolerant patients who show serious akathisia and parkinsonian side effects even on low doses of low-potency neuroleptics. Such patients are often noncompliant and may become more compliant on clozapine [2]. Clozapine should be considered in patients with moderate to severe tardive dyskinesia (TD), especially if it is progressive or not blocked by typical neuroleptic drugs. Between 50 and 75% of such patients will respond to clozapine with a decrease in abnormal movements [8–10]. Because clozapine has not been shown to produce TD or dystonia, it may be useful to consider clozapine when the first signs of TD or dystonia emerge, especially in groups with high risk for TD, e.g. older women.

Starting Clozapine Treatment

Whenever possible, clozapine should initially be used as monotherapy in treatment-resistant schizophrenic patients. Monotherapy will make titration easier but it also has a theoretical rationale. Neuroleptic drugs occupy D_2 receptors and the most coherent theories of clozapine's advantages over typical neuroleptic drugs are based upon its ability to produce less D_2 receptor blockade [11]. Patients should be withdrawn from all other psychotropic drugs, especially carbamazepine and including neuroleptics. The former may cause hematologic complications. It is possible, although unproven, that a synergistic effect with clozapine to cause bone marrow toxicity might occur. If antiepileptic medication is needed, valproate is the drug of choice. The latter may interfere with, rather than augment the action of clozapine. To accomplish this, hospitalization may be needed. But with a strong outpatient program and family cooperation, a drug-free start is even possible on an outpatient basis.

Clozapine should be started at a low dose, e.g., 25 mg/day, and the dose built up slowly to 300–500 mg/day by increments of 25 mg every other day. These may be larger if tachycardia, hypotension or sedation do not cause excessive problems. If patients do require a neuroleptic when they begin clozapine, it should be a high-potency agent, given at a low dose (e.g., haloperidol 5 mg b.i.d.) and it should be tapered after the dose of clozapine reaches 200–300 mg/day. Clozapine plasma levels may be clinically useful [12]. The steady state is reached after 5–7 days. Levels 12 h after the last dose should be ≥ 350 ng/ml.

Psychosocial Treatments

Treatment-resistant schizophrenic patients often require extensive psychosocial support to maximize the benefits of clozapine. They may include: education, group therapy, psychosocial skill training, job training, activity counseling, etc. Similarly, families of these patients may need extensive help to reestablish better relations with their previously refractory family member. Assistance to find paid or volunteer work is often required for deinstitutionalized chronic patients. Treatment systems which provide long-term custodial care should be less necessary as clozapine becomes widely used because it facilitates discharge and diminishes the need for hospitalization. Resources will need to be diverted to outpatient programs to accommodate the treatment-resistant cases.

Augmentation of Clozapine

If response to clozapine alone is not adequate by 6 months, it is possible that addition of a neuroleptic drug, an antidepressant, lithium or

Table 1. Dimensions of outcome

1	Psychopathology	4	Rehospitalization
	Cognitive function	5	Suicide
2	EPS (TD)	6	Compliance
3	Social function	7	Family burden
	Independent living	8	Cost of illness and treatment
	Work function		
	Quality of life		

electroconvulsive treatment may be helpful. Each such trial should be done individually for up to 6 weeks. It is desirable to utilize rating scales to monitor progress in such cases. The choice among these treatments might be based upon the major symptomatology which has not yielded to clozapine.

Outcome: A Multidimensional Perspective

A comprehensive assessment of outcome of treatment with clozapine should consider the dimensions listed in table 1. The importance of specific consideration of each of these outcome measures is supported by the independence of many of them [13–15], as well as by the need to make risk-benefit assessments for individual measures. The latter issue is particularly important with regard to clozapine because of the 1–2% incidence of granulocytopenia or agranulocytosis associated with its use [16]. This risk must be considered in relation to the potential benefits of clozapine as will be emphasized in the conclusion of this article.

Effect of Psychopathology

The short-term benefits of clozapine in treatment-resistant patients were established in the study of Kane et al. [1]. Improvement in positive and negative symptoms and ward behavior in 30% of treatment-resistant patients over a 6-week period was found. This was replicated by Meltzer et al. [7]. The 30% figure is generally supported by other investigators. However, Verhoeven et al. [17] recently reported that 33 of 57 (58%) treatment-resistant schizophrenic patients had at least a 50% decrease in the Brief Psychiatric Rating Scale (BPRS) [18] total score over a 18-week period. This may be misleading because it is possible that a 0–6 rather than a 1–7 scoring system was used. It is also possible that many neuroleptic-intolerant patients who were not neuroleptic-resistant were included in this sample. These patients would be expected to have a better response to clozapine. There have been several retrospective long-term studies of the outcome of clozapine treatment [19, 20] which indicate clozapine reduces symptoms, decreases hospitalization and enables some patients to work. In a prospective study, Meltzer et al. [7] reported the results of an open trial of 51 patients who had completed at least 6 weeks and up to 35 months (mean 10.3 ± 8.1 SD months) of treatment with clozapine. There was a significant decrease in BPRS at 6 weeks, 3, 6, 9 and 12 months, with a progressive improvement over time.

We have now analyzed data on 132 consecutive patients with a DSM-III-R diagnosis of schizophrenia who were treatment-resistant by the criteria of failure to respond to at least 2, and in almost all cases, 3 or more trials of typical neuroleptic drugs. These patients included 85 patients previously described by Meltzer et al. [4, 7]. Of the 132 patients, 107 (81.1%) were inpatients at University Hospitals of Cleveland and 25 (18.9%) were treated at the Cleveland Veterans Administration Hospital. Many of the University Hospitals patients were medically indigent.

Of these 132 patients, 90 (68.2%) remained on clozapine at the end of 12 months. Thirteen patients (9.9%) dropped out because of adverse reactions, including 4 (3.0%) cases of granulocytopenia or agranulocytosis. Eight (6.1%) patients were dropped because of lack of efficacy. Nineteen (14.4%) were discontinued because of noncompliance. A comparison of the 90 patients who remained on clozapine for 12 months with the 42 who did not is given in table 2. There were no significant differences in age, sex, age of onset, number of previous hospitalizations, baseline BPRS Total, Positive or Negative Subscale scores, Global Assessment Scale (GAS) or Quality of Life Scale (QLS) scores [21].

The total BPRS score, BPRS Positive Symptoms, Schedule for Affective Disorders-Change (SADS-C) [22], Negative Symptoms and Disorganization Symptoms for 63 of the 90 patients who remained on clozapine for ≥12 months are given in figure 1. This subgroup was chosen because complete BPRS data was available for them at baseline, 6 weeks, 3, 6, 9 and 12 months. Significant improvement was noted in all four measures at each time point during the 12-month period. Because this was an open trial

Table 2. Comparison of patients completing 12 months clozapine treatment and dropouts

Measure	Completers	Dropouts
Age	33.3 ± 9.3 (90)	34.9 ± 8.9 (42)
Sex	64 M, 26 F	24 M, 18 F
Age of onset	20.0 ± 4.7 (74)	20.2 ± 6.0 (36)
Number of hospitalizations	8.3 ± 6.5 (71)	9.6 ± 9.9 (30)
BPRS total	48.8 ± 13.0 (90)	48.9 ± 15.1 (42)
SADS-C positive	3.9 ± 2.5 (77)	3.7 ± 2.7 (27)
SADS-C negative	4.8 ± 3.5 (84)	4.9 ± 4.0 (37)
SADS-C disorganization	3.8 ± 3.7 (82)	3.2 ± 3.5 (34)
Global assessment	31.3 ± 8.9 (84)	32.5 ± 11.4 (37)
Quality of life	32.4 ± 20.6 (78)	32.5 ± 16.4 (211)

Number of patients in parentheses.

which was designed, in part, to determine when clinical response occurred, and because multiple outcome measures were considered relevant, patients who showed little or no change in psychopathology were treated with clozapine beyond the conventional 6-week trial period. The data in figure 1 indicate that the major improvement in psychopathology in the group as a whole occurred during the first 6-week period. Forty-five of the 132 (34.1%) patients had a decrease in BPRS total score ≥20% at 6 weeks (34.1%). This is similar to the responder rate at 6 weeks in the study of Kane et al. [1]. However, the best response to clozapine by 12 months often occurred after 6 weeks. Thirty-five of the 132 patients (26.5%) showed improvement in their BPRS total score ≥40%; of these 35, 24 (68/6%) first reached this level by 3 months.

In agreement with our previous report that response to clozapine treatment is often delayed, we examined the time in which a 20% decrease in BPRS was first recorded (table 3). It can be seen that 78 of the 132 (59.1%) patients had a decrease in BPRS ≥20%. Of these 78, 45 (57.7%) responded by this criterion at 6 weeks. Seventeen of the 78 responders did so first at 3 months. Another 7 and 3 did so at 9 and 12 months. This provides the rationale for prolonged trials (at least 6 months) with clozapine before deciding that clozapine does not provide any advantage over typical neuroleptic drugs for a given treatment-resistant patient.

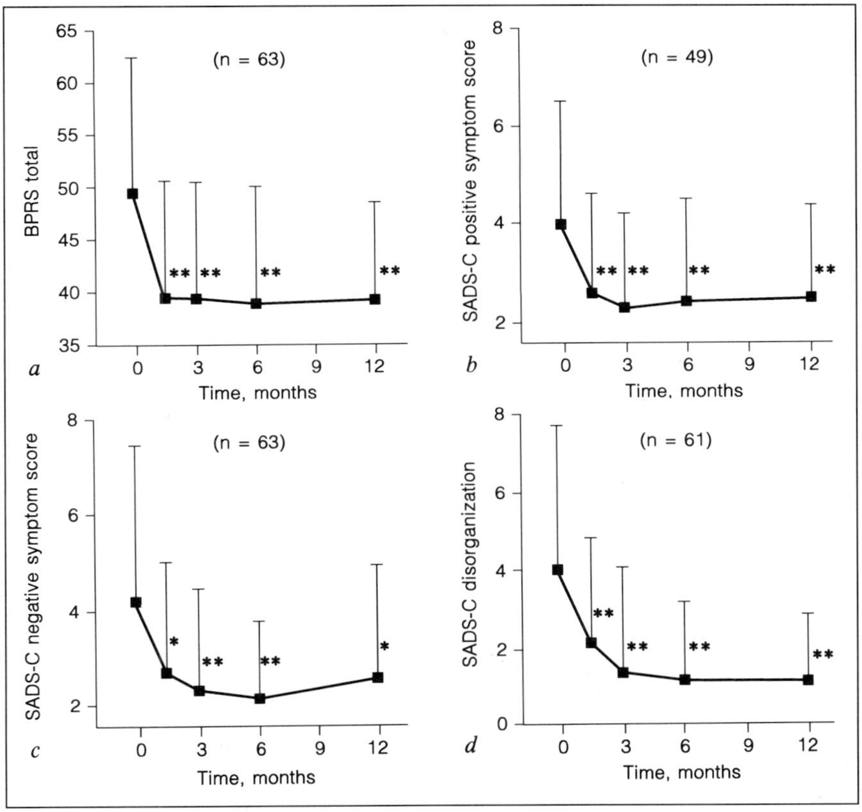

Fig. 1. BPRS total score (*a*); BPRS Positive Symptoms (*b*); SADS-C Negative Symptoms (*c*), and SADS-C Disorganization Symptom (*d*) in 54 treatment-resistant schizophrenic patients who remained on clozapine ≥ 24 months. The number of patients with available data at each time period is given in parentheses. Repeated measures analysis of variance of the data for subjects with no missing time points showed significant time effects (BPRS total, $p = 0.0001$; SADS-C positive, $p = 0.03$; SADS-C negative, $p = 0.0001$; SADS-C disorganization, $p = 0.003$). * $p < 0.01$; ** $p < 0.0001$; error bars are standard deviation.

A factor analysis of the SADS-C items which relate to psychosis identified a Positive Symptom factor (severity of delusions and hallucinations), a Negative Symptom factor, consisting of the following items: loss of interest, depressed appearance, slowed body movement and slowed speech; and a separate Disorganization factor, consisting of loose associations, poverty

Table 3. Time of first response to clozapine by BPRS criterion

Time of first response, months	Patients		Cumulative percent
	n	%	
1½	45	57.7	57.7
3	17	21.8	79.5
6	6	7.7	87.2
9	7	9.0	96.2
12	3	3.9	100.0

of thought content, incoherence and inappropriate affect [6]. Improvement in the SADS Disorganization factor was particularly impressive. Although the SADS Negative Symptom factor includes items related to avolition and anergia only, it correlates highly with the sum of the five global ratings of the Scale for the Assessment of Negative Symptoms [23] ratings (rho = 0.65, n = 203, p = 0.0001).

Positive and negative symptoms on the BPRS and SADS-C were not significantly correlated at baseline and 12 months [data not presented]. Patients often improved in only positive or negative symptoms. Six of the 12 patients who had predominantly negative symptoms with relatively low levels of positive symptoms had a 50% or greater decrease in SADS-C negative symptoms during the course of 12 months treatment. This indicates that clozapine may be an effective treatment for those patients who phenomenologically resemble the type II syndrome of Crow [24].

Cognitive Function

Cognitive abnormalities are a critical aspect of schizophrenia that is now coming into particular prominence as an important deficit that impacts an overall outcome [25]. These abnormalities include deficits in attention, working memory, semantic memory, and executive function [25–27]. Neuroleptic drugs may improve attentional deficits but memory measures are fairly resistant to drug action. This has sometimes been attributed to the anticholinergic properties of some of these drugs [28, 29] but it may also be due to structural abnormalities in the hippocampus and frontal cortex.

We examined the effects of clozapine on four types of neuropsychological measures in 25 treatment-resistant schizophrenic patients [30]. Significant improvement was noted in the Controlled Word Association tests of Semantic Memory after 6 weeks and 6 months of treatment, and in the Immediate and Delayed Recall test of Secondary Memory only at 6 months. Category Instance Generation, a measure of Semantic Memory, also improved at 6 months only. There was no overall improvement in tests of Executive Function (Wisconsin Card Sort) or Attention/Short-Term Memory (as assessed by the Digit Symbol Test and Consonant Trigram Test). However, for all tests there were some patients who improved while others showed no change or even poorer performance. These changes in cognitive function were independent of changes in psychopathology.

The improvement in some aspects of cognitive function reported here must be verified in an independent sample. In any event, the changes still leave most of the patients still impaired compared to normal controls in most of the areas tested. It may be that the ability of clozapine to improve cognitive deficits may be greater if treatment with it is initiated earlier in the course of illness. The duration of illness in the patients in this study averaged 12–14 years.

Extrapyramidal Symptoms (TD)

Neuroleptic drugs, parkinsonain side effects, e.g., rigidity, tremor, dystonias, masked facies, etc. are a nearly constant concomitant of neuroleptic drug treatment, especially at higher dosages [31]. Treatment-resistant schizophrenic patients are especially likely to receive higher dosages. While some neuroleptics such as thioridazine produce fewer extrapyramidal symptoms (EPS), even thioridazine is unlikely to be entirely free of EPS in vulnerable individuals. Parkinsonian side symptoms, especially akathisia, have been suggested to be the major reason for noncompliance with antipsychotic drug treatment [32]. Clozapine produces much less EPS than standard neuroleptic drugs [33, 34].

Clozapine produces significantly less TD than other antipsychotic drugs [10, 33]. Two patients have been reported to develop mild abnormal involuntary movements after many years on clozapine, but these had questionable movement disorder prior to the period of observation [Kane, personal commun.]. In view of the reports of tardive-like movement disorders in schizophrenic patients before the neuroleptic era, it is possible that these cases do not represent the effect of clozapine on TD. The advantage of clozapine with regard to EPS and lack of causation of TD must be considered in evaluating risk-benefit issues with the use of this drug. It is important for compliance and also for the general sense of well-being that patients experience few EPS and are virtually free of the fear of developing or exacerbating TD.

Quality of Life

Quality of life is a concept that is receiving increasing attention throughout medicine. It is particularly relevant to chronic illnesses, such as schizophrenia in which the illness can transform the lives of the afflicted people and become the determining factor of how they live and work, how they relate to significant others, and what, if anything, in the way of pleasurable, satisfying daily activity, they can engage in.

We have now studied 59 patients over a 12-month period. Baseline QLS scores increased from 36.8 ± S.D. 18.7 to 72.5 ± 24.1. Instrumental Role Function showed as much improvement as Interpersonal Role and Intrapsychic Functions [4, 35].

The improvement in quality of life we have observed with clozapine has made it possible for some chronically disabled individuals who were leading marginal social lives, in and out of hospitals, but usually in some type of community residence, to resume a level of functioning that would be considered within the low normal range because it includes stable interpersonal relationships, varied social activities, perhaps a limited job, and living with their families.

Rehospitalization

We previously reported the hospitalization data for 85 patients who had begun on clozapine at least 12 months before follow-up [4]. Of these 85, 29 (34.1%) were rehospitalized during the 12 months after beginning clozapine compared to 73 (85.8%) in the 12 months before clozapine, excluding the index admission ($p < 0.001$). The total number of admissions to hospital for all 85 patients decreased from 118 to 33, a 72.0% decrease. Of the 31 patients who discontinued clozapine, 13 (42%) required hospitalization. Of the 54 on clozapine for 12 months, 16 (29.6%)

required hospitalization. The 16 patients who needed hospitalization while on clozapine, and who remained on it during and after hospitalization, usually recovered within 2–3 weeks and resumed outpatient treatment.

The low level of hospitalization was achieved with the aid of an outpatient program that included many elements common to contemporary treatment programs: family and patients education, various types of groups, social skills training, vocational counseling, etc. The significant decrease in rehospitalization in this group may represent the persistent benefits of clozapine or psychosocial treatment, or both.

Compliance

Despite the burden of weekly white blood cell counts, compliance with clozapine is usually excellent. The incidence of noncompliance in our sample was 19 of 132 (14.4%). The noncompliance rate in chronic schizophrenic patients who receive typical antipsychotic drugs orally in our experience has been much higher. The low EPS, improvement in three dimensions of psychopathology, education about the long-term advantages of clozapine treatment, and mandated weekly contact with the clinic are the major factors in improved compliance. A significant number of treatment-resistant patients refuse to begin clozapine treatment because of fear of blood drawing or skepticism about the benefits of clozapine in relation to its risks. Intensive effort in educating the patient and the family often is successful in gaining the consent of these individuals. It has been our experience that providing an opportunity for candidates for clozapine treatment to meet patients on clozapine has been helpful in overcoming resistance to clozapine treatment. The greater compliance with clozapine than with typical neuroleptic drugs no doubt contributes to the low hospitalization rate, the continuing decrease in psychopathology and improvement in quality of life.

Cost of Illness and Treatment

We have examined the cost of treatment for 37 University Hospitals patients in the 2 years before clozapine and the 2 years after treatment was begun [Meltzer et al., in preparation]. We also found that the median savings with clozapine over a 2-year period was $31,300/patient, despite including the cost of clozapine, weekly monitoring of the white blood count and case management at $9,000/year. This saving was mainly due to a dramatic saving in the cost of hospitalization (median $35,435 for 2 years) together with some decrease in ancillary costs of illness to the family, not from discharge of chronically hospitalized patients.

Parasuicide Events

Suicide is a not infrequent outcome in schizophrenia. Up to 7–10% of patients with schizophrenia commit suicide [36]. There is some evidence that treatment-resistant patients suicide more frequently than nontreatment-resistant patients [37–39]. We observed a decrease in the number of patients who made suicide attempts with a low or high probability of succeeding while on clozapine compared to the comparable period before initiating clozapine treatment. This was paralleled by a significant decrease in depressive symptoms [Meltzer et al., in preparation].

Risk-Benefit Considerations

The major risk of clozapine is serious infectious illness and possible death due to agranulocytosis. As previously mentioned, agranulocytosis occurs in at most 1–2% of patients with vigorous monitoring, death occurs in 2–3 % of these patients. The major period of risk for agranulocytosis is during the first 4–5 months. While it can occur thereafter, current evidence suggests that the risk after this period is greatly reduced. Thus, after the initial 4-month period, the risk-benefit ratio changes appreciably. Other serious side effects of clozapine such as seizures (4–6%), tachycardia and respiratory difficulties have proven manageable in clinical practice.

The benefits of clozapine are of major importance to the outcome of schizophrenia. Most of these have been described in this article and elsewhere [4] and need not be repeated here. There is some evidence that the advantages of clozapine in diminishing positive and negative symptoms hold even for nontreatment-resistant schizophrenic patients [40, 41; Meltzer et al., unpubl. data]. The almost absent risk of developing TD with clozapine is also of great importance in assessing its risk/benefit ratio. The ability to decrease the risk of suicide is also of importance in this regard.

For these reasons, clozapine is being used to treat schizophrenic patients who show little or no treatment resistance by clinicians in many countries who have used it with success in more definitively treatment-resistant schizophrenia. Further study to definitely establish the benefits of clozapine in schizophrenic patients showing a much wider range of responsiveness to typical neuroleptic drugs is definitely in order before this practice becomes widespread. As experience is gained in using clozapine safely, and especially if means of preventing agranulocytosis or identifying patients at risk for this side effect are developed, a much wider use of clozapine in schizophrenia would seem to be possible.

Conclusion

Experience with clozapine in treatment-resistant schizophrenia highlights the utility of multidimensional outcome measures and permitting sufficient time on clozapine for these benefits to emerge. Improvement in psychopathology should be only one of a group of outcome measures. It may be highly predictive of other benefits, but improvement in social function, decreased hospitalization, better compliance and fewer EPS may be of equal or greater importance in patients. Moreover, these may even be independent of improvement in psychopathology. Only 34% of a group of 132 highly treatment-resistant schizophrenic patients showed improvement at 6 weeks by the single criterion of a decrease in total BPRS score $\geq 20\%$. However, by 6–12 months improvement in other outcome measures was present in another 34% of the sample.

Acknowledgements

The research reported was supported in part by USPHS MH 41684, the Department of Veterans Affairs, and grants from the John Pascal Sawyer and Elizabeth Severance Prentiss Foundation. H.Y.M. is the recipient of a USPHS Research Career Scientist Award MH 47808. The assistance of Mr. Phil Cola and Ms. Lynne Way in data collection and analyses, and the secretarial assistance of Ms. Lee Mason is greatly appreciated.

References

1 Kane J, Honigfeld G, Singer J, Meltzer HY, the Clozapril Collaborative Study Group: Clozapine for the treatment-resistant schizophrenic: A double-blind comparison with chlorpromazine. Arch Gen Psychiatry 1988;45:789–796.

2 Marder SR, Van Putten T: Who should receive clozapine? Arch Gen Psychiatry 1988;45:865–867.

3 Ereshefsky L, Watanabe MD, Tran-Johnson TK: Clozapine: An atypical antipsychotic agent. Clin Pharmacol 1989;8:691–709.

4 Meltzer HY: Dimensions of outcome with clozapine. Br J Psychiatry 1992; 160(suppl 17):46–53.

5 Meltzer HY: Defining treatment refractoriness in schizophrenia. Schizophr Bull 1991;17:563–566.

6 Thompson P, Meltzer HY: Positive, negative and disorganization factor from the Schedule of Affective Disorders and Schizophrenia-Change scale. Br J Psychiatry, in press.

7 Meltzer HY, Bastani B, Kwon KY; Ramirez LF, Burnett S, Sharpe J: A prospective study of clozapine in treatment-resistant patients: I: Preliminary report. Psychopharmacology 1989;99:S68–S72.

8 Caine ED, Polinsky RJ, Kartzinel R, Ebert MH: The trial use of clozapine for abnormal involuntary movement disorders. Am J Psychiatry 1979;136:317–320.

9 Meltzer HY, Luchins DJ: Effect of clozapine in severe tardive dyskinesia: A case report. J Clin Psychopharmacol 1984;4:286–287.

10 Lieberman J, Johns C, Cooper T, Pollack S, Kane J: Clozapine pharmacology and tardive dyskinesia. Psychopharmacology 1989;99:S54–S59.

11 Meltzer HY: The mechanism of action of novel antipsychotic drugs. Schizophr Bull 1991;17:263–287.

12 Perry PJ, Miller DD, Arnett SV, Cardoret RJ: Clozapine and norclozapine plasma concentrations and clinical response of treatment-refractory schizophrenia patients. Am J Psychiatry 1991;148:231–235.

13 Strauss JS, Carpenter WT: The prediction of outcome in schizophrenia: I. dharacteristics of outcome. Arch Gen Psychiatry 1972;27:734–746.

14 Strauss JS, Carpenter WT: The prognosis of schizophrenia: Rationale for a multidimensional concept. Schizophr Bull 1978;4:56–67.

15 Strauss JS, Carpenter WT, Bartko JJ: The diagnosis and understanding of schizophrenia: III. Speculations on the processes that underlie schizophrenia symptoms and signs. Schizophr Bull 1974;1:61–69.

16 Krupp P, Barnes P: Leponex-associated granulocytopenia: A review of the situation. Psychopharmcology 1989;99:S118–S121.

17 Verhoeven WMA, Doesburg WH, Snoej R, Rutgers AJMP, van Dongen PHM: Efficacy of clozapine in treatment-resistant psychosis and neuroleptic sensitivity: Result of the Dutch open multi-center project. Eur Psychiatry 1992;7:77–84.

18 Overall JE, Gorham DR: Brief Psychiatric Rating Scale. Psychol Rep 1962;10:799–812.

19 Polvsen JU, Noring U, Fog R, Gerlach J: Tolerability of therapeutic effect of clozapine: A retrospective investigation of 216 patients treated with clozapine for up to 12 years. Acta Psychiatry Scand 1985;71:176–195.

20 Lindström LH: The effect of long-term treatment with clozapine in schizophrenia: A retrospective study in 96 patients treatment with clozapine for up to 13 years. Acta Psychiatr Scand 1988;77:524–529.

21 Heinrichs DW, Hanlon TE, Carpenter WT Jr: The Quality of Life Scale: An instrument for rating the schizophrenic deficit syndrome. Schizophr Bull 1984;10:388–398.

22 Endicott J, Spitzer RL: A diagnostic interview: The schedule for affective disorders and schizophrenia. Arch Gen Psychiatry 1978;35:837–844.

23 Andreasen NC: Negative symptoms in schizophrenia: Definition and reliability. Arch Gen Psychiatry 1982;39:784–788.
24 Crow TJ: Molecular pathology of schizophrenia: More than one disease process? Br Med J 1980;280:66–68.
25 Kenny JT, Meltzer HY: Attention and higher cortical functions in schizophrenia on semantic memory. J Neuropsychiatry Clin Neurosci 1991;3:269–275.
26 Chapman LJ: Recent advances in the study of schizophrenia cognition. Special Report: Schizophrenia. Bethesda, US Department of Health and Human Services, 1980, pp 69–81.
27 Seidman LJ: Schizophrenic and brain dysfunction: An integration of recent neurodiagnostic findings. Psychol Bull 1983;4:195–238.
28 Spohn HE, Strauss ME: Relation of neuroleptic and anticholinergic medication to cognitive functions in schizophrenia. J Abnorm Psychol 1989;98:367–380.
29 Perlick D, Stastny P, Katz U, Mayer M, Mattis S: Memory deficits and anticholinergic levels in chronic schizophrenia. Am J Psychiatry 1986;143:230–232.
30 Hagger C, Friedman L, Ubogy D, Meltzer HY: Effect of clozapine on neuropsychological functions. Biol Psychiatry, submitted.
31 Keepers GA, Clappison VJ, Casey DE: Initial anticholinergic prophylaxis for neuroleptic-induced extrapyramidal syndromes. Arch Gen Psychiatry 1983;40:1113–1117.
32 Van Putten T, Marder SR: Behavioral toxicity of antipsychotic drugs. J Clin Psychiatry 1974;48:13–19.
33 Casey DE: Clozapine: Neuroleptic-induced EPS and tardive dyskinesia. Psychopharmacol Bull 1989;99:S47–S53.
34 Safferman A, Lieberman JA, Kane JM, Szymanski S, Kinon B: Update on clinical efficacy and side effects of clozapine. Schizophr Bull 1991;17:247–262.
35 Meltzer HY, Burnett S, Bastani B, Ramirez LF: Effect of six months of clozapine treatment on the quality of life of chronic schizophrenic patients. Hosp Community Psychiatry 1990;41:892–897.
36 Roy A: Suicide in schizophrenia; in Roy A (ed): Suicide. Baltimore, Williams & Wilkins, 1986, pp 95–112.
37 Yarden PE: Observations on suicide in chronic schizophrenics. Compr Psychiatry 1974;15:325–333.
38 Roy A, Mazonson A, Pickar D: Attempted suicide in chronic schizophrenia. Br J Psychiatry 1984;144:303–306.
39 Drake R, Gates C, Cotton P, Whitaker A: Suicide among schizophrenia. Who is at risk? J Nerv Ment Dis 1984;172:613–617.
40 Claghorn J, Honigfeld G, Abuzzahab FS, Wang R, Steinbock R, Tuason V, Klerman G: The risks and benefits of clozapine versus chlorpromazine. J Clin Psychopharmacol 1987;7:377–384.
41 Baldessarini RJ, Frankenburg FR: Clozapine: A novel antipsychotic agent. N Engl J Med 1992;324:746–754.

Herbert Y. Meltzer, MD, Bond Professor of Psychiatry,
University Hospitals of Cleveland, Hanna Pavilion, Room B-68,
2040 Abington Road, Cleveland, OH 44106-5000 (USA)

Brunello N, Mendlewicz J, Racagni G (eds): New Generation of Antipsychotic Drugs: Novel Mechanisms of Action. Int Acad Biomed Drug Res. Basel, Karger, 1993, vol 4, pp 118–129

Future Strategies in the Discovery of New Antipsychotic Agents: Focus on Dopamine-Glutamate Interactions

Arvid Carlsson, Anders Svensson, Maria L. Carlsson

Department of Pharmacology, University of Göteborg, Sweden

Introduction

The heuristic value of the dopamine hypothesis of schizophrenia can hardly be denied. However, some important caveats should be considered. The hypothesis rests almost entirely on pharmacological data. No disturbance in dopaminergic functions in the brains of schizophrenic patients have so far been demonstrated beyond doubt. The response to treatment with antidopaminergic drugs is often not entirely satisfactory. Dopaminergic agonists can primarily mimic the paranoid form of schizophrenia. Non-paranoid schizophrenia can be mimicked more faithfully by phencyclidine (PCP, 'angel dust') [1, 2], an antagonist on the glutamatergic NMDA receptor [3]. A deficient glutamatergic function in schizophrenia was proposed by Kornhuber et al. [4], and Kim et al. [5].

The dopamine hypothesis of schizophrenia can best explain the 'positive' symptoms (hyperarousal, hallucinations etc.) and may thus rather be looked upon as a hypothesis of psychosis in general. The 'negative' symptoms (affective flattening, poverty of speech) [6] are rather suggestive of a dopaminergic hypofunction. A comprehensive hypothesis must be able to offer an explanation to the fact that 'positive' and 'negative' symptoms can occur simultaneously and independently. In this review we shall try to explain the 'positive' and 'negative' symptoms of schizophrenia against the background of recent neuroanatomical and pharmacological data, especially dealing with the ventral (limbic) striatum and its connections. Some therapeutic implications will then be discussed.

The Possible Involvement of Different Neuronal Circuitries in the Basal Ganglia in 'Positive' and 'Negative' Symptoms of Schizophrenia

One major input to the striatum comes from the cerebral cortex (fig. 1), that is, from the neocortex to the dorsal striatum and from the allocortex (and the prefrontal cortex) to the ventral (limbic) striatum. From the dorsal and ventral striatum major projections go to different thalamic nuclei via the dorsal and ventral pallidum, respectively. The loop is closed by projections from the thalamus back of the cortex.

The inhibitory functions of the striatum on behavior is evident from the fact that removal of the excitatory glutamatergic tone in ventral striatum causes behavioral stimulation even in the virtually complete absence of dopamine. Glutamate and dopamine can thus antagonize each other in the striatum (fig. 2) [7–10]. The inhibitory influence of the striatum on the thalamus may be exerted via the so-called 'indirect' striato-pallido-thalamic pathways, which contain chains of three gabaergic neurons (and in addition one glutamatergic neuron originating in the subthalamic nucleus (STN) and projecting to the medial segment of the globus pallidus, see fig. 1). More recently we have become aware of another gabaergic, 'indirect' pathway, consisting of a 3-neuron chain from the striatum to the thalamus, with the third gabaergic neuron located in the reticular nucleus of thalamus [11–13]. The latter nucleus is known to control most of the other thalamic nuclei. Taken together, these 'indirect' 3-neuron gebaergic pathways should be able to mediate an inibitory impact on the thalamus, resulting in inhibition via the motor feedback loops as well as a reduced thalamic relaying of the sensory input to the cortex.

The 'direct' striato-pallido-thalamic pathways, which contain only two gabaergic neurons [14, 15], appear to serve a disinhibitory behavioral function, and mediate positive feedback. The dopaminergic input to those striatal neurons which are involved in the 'direct' pathways, appear to be excitatory on these neurons [16], and dopamine should thus be behaviorally stimulating also via the 'direct' pathways. Since dopamine seems to be excitatory on the 'direct' and inhibitory on the 'indirect' pathways, it should regulate the balance between positive and negative feedback. A low dopaminergic tone would favor negative and a high tone positive feedback (fig. 3).

When we examined the interaction between glutamate and dopamine at the postsynaptic level (in monoamine-depleted mice), we observed that D_1 and D_2 agonists behaved differently. Whereas the behavioral actions of

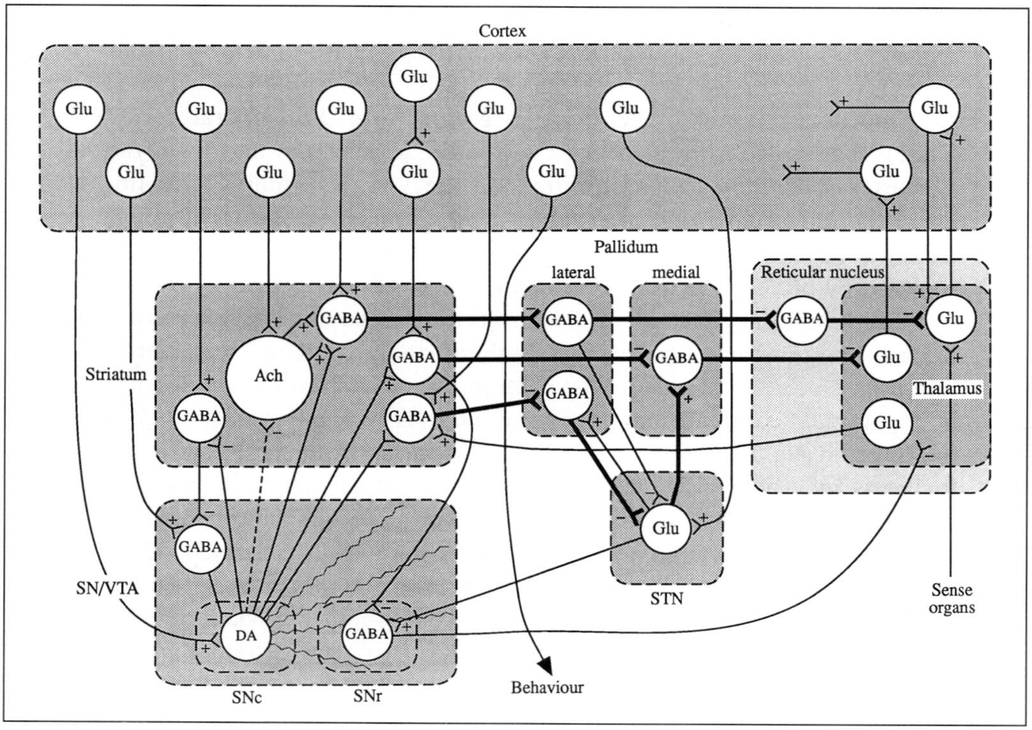

Fig. 1. Neurocircuitries of the basal ganglia. The pathways drawn with thick lines indicate some connections between the striatum and the thalamus, which are discussed in some detail in the present paper. The top and bottom pathways drawn with thick lines contain three gabaergic neurons and are referred to as 'indirect' pathways. The pathway in between contains two gabaergic neurons and is referred to as 'direct'. SN = Substantia nigra; VTA = ventral tegmental area; Glu = glutamate; ACh = acetylcholine; DA = dopamine.

the D_1 agonist SKF 38393 and the mixed D_1/D_2 agonist apomorphine, given in threshold doses, were potentiated by the NMDA antagonist dizocilpine [8, 9], the actions of D_2 agonists were antagonized by this agent (fig. 4) [17]. Using somewhat different experimental models Morelli and Di Chiara [18] and Goodwin et al. [19] have reported similar results. Thus, the responsiveness of the D_1-receptor is apparently suppressed by the glutamatergic tone, whereas D_2-receptors seem to cooperate with the glutamatergic system to enhance behavioral stimulation. Glutamate can either

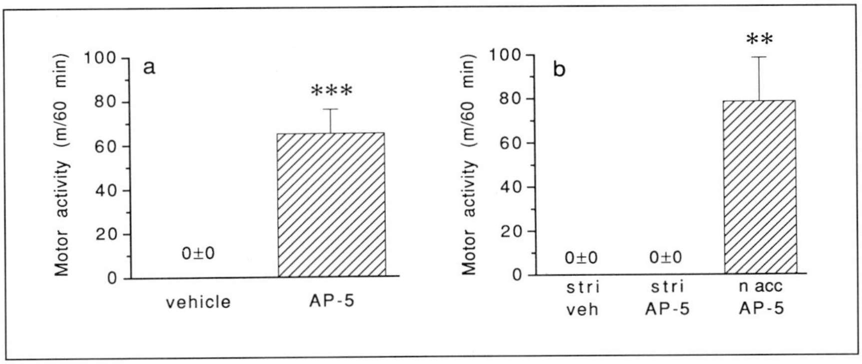

Fig. 2. In the experiments shown in this figure the mice were monoamine-depleted by means of pretreatment with reserpine (10 mg/kg) 18 h and α-MT (500 mg/kg) 2 h before the registration of locomotor activity commenced. *a* Effects of AP-5 injected into the nucleus accumbens. 10 min following the injection of AP-5 (5 μg) or vehicle the animals were placed in the circular tracks and locomotor activity was registered during 60 min. Shown are the means ± SEM, n = 6–8. *** p = 0.0006 vs. vehicle. *b* Effects of AP-5 injected into the nucleus accumbens or dorsal striatum. 10 min following the injection of AP-5 (5 μg) or vehicle the animals were placed in the circular tracks and locomotor activity was registered during 60 min. Shown are the means ± SEM, n = 4–6. ** p = 0.0039 vs. AP-5 injected into the dorsal striatum.

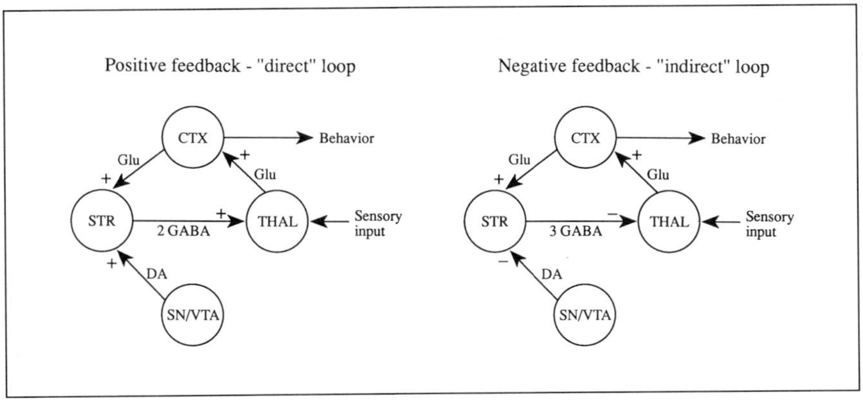

Fig. 3. Schematic diagram illustrating positive and negative feedback loops in the striatum, mediated via 'direct' and 'indirect' pathways, respectively. Note that in this somewhat hypothetical scheme dopamine is excitatory on the 'direct' and inhibitory on the 'indirect' pathways, as first suggested by Penney and Young [14].

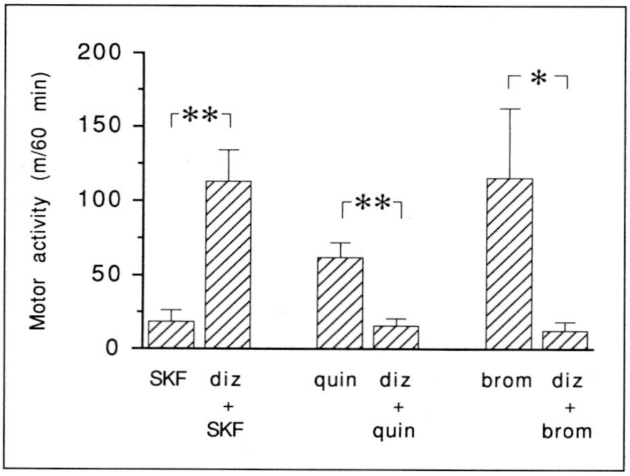

Fig. 4. Effects of dizocilpine (diz) in combination with either SKF 38393 or quinpirole (quin; experiment 1) or bromocriptine (brom; experiment 2) on locomotor activity in monoamine-depleted mice. Bromocriptine (10 mg/kg) was given 150 min, dizocilpine (1 mg/kg) 60 min, and quinpirole (1 mg/kg) and SKF 38393 (8 mg/kg) immediately before the locomotor recording commenced. Locomotor activity was recorded in the circular tracks for 60 min. Shown are the means ± SEM, n = 5–6. ** $p < 0.01$, * $p = 0.01$.

inhibit or stimulate behavior, and the outcome seems to depend to some extent on whether it interacts with D_1- or D_2-receptors. However, the behavioral baseline level of activity was also important for the outcome. Thus, with a low baseline activity the ability of an NMDA receptor antagonist to enhance locomotor activity is especially strong and may show up, though only weakly, even in conjunction with a D_2-receptor agonist. With a high baseline activity an NMDA antagonist tends to be inhibitory even in conjunction with a D_1-receptor agonist. Especially striking was the dependence on baseline activity in experiments with the mixed D_1/D_2 agonist apomorphine, in that dizocilpine potentiated the behavioral stimulation induced by a threshold dose of apomorphine but antagonized the overt stimulation caused by a higher dose of this drug [17].

Perhaps even more striking are some more recent observations in our laboratory [20], dealing with the effects of unilateral administration of the competitive NMDA antagonist AP-5 into the nucleus accumbens of the mouse. In the monoamine-depleted mouse this results in behavioral activation and contralateral rotation (fig. 5). In mice with intact monoaminer-

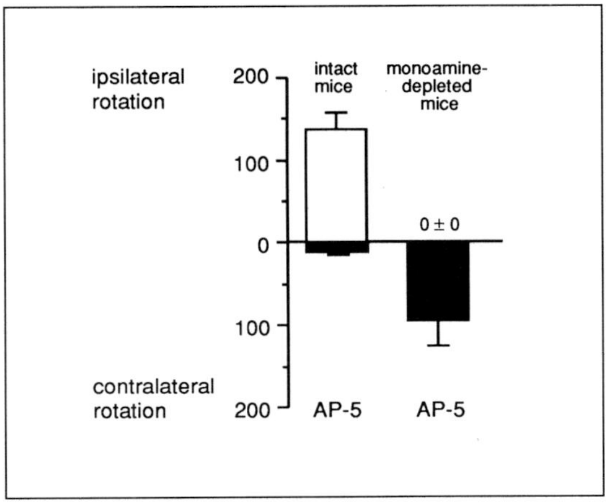

Fig. 5. Rotatory behavior of mice following unilateral injection of the competitive NMDA receptor antagonist AP-5 into the nucleus accumbens of mice. Monoamine-depleted mice were pretreated with reserpine (10 mg/kg) 18 h and α-MT (500 mg/kg) 2 h before recording of rotatory behaviour. Immediately after the injection of AP-5 (5 μg) the animals were videotaped for 20 min and the number of complete rotations was counted manually. Shown are the means ± SEM, n = 6–8.

gic systems this treatment induces predominantly ipsilateral rotation. These observations are in agreement with our model, because the 'indirect' pathways and negative feedback should predominate in monamine-depleted animals. Here the unilateral loss of glutamatergic tone should reduce the striatal inhibition ipsilaterally, leading to contralateral rotation. In mice with an intact dopaminergic system the 'direct' pathways and positive feedback should predominate, and thus unilateral inhibition of glutamatergic tone should cause loss of positive feedback ipsilaterally and ipsilateral rotation. The systemic administration of the dopamine D_2-receptor agonist quinpirole or the mixed D_1/D_2 agonist apomorphine to monoamine-depleted mice with unilateral reduction of glutamatergic tone, induced by AP-5, caused a switch of rotation from contralateral to ipsilateral. Thus an increased D_2-receptor-mediated dopaminergic tone appears to switch the system from predominantly negative to positive feedback. On the other hand, the dopamine D_1-receptor agonist SKF 38393, injected systematically to monoamine-depleted mice with AP-5 given unilaterally into nucleus accumbens, caused increased contralateral rotation. These experiments strongly support the view that glutamate serves a dual function in the nucleus accumbens and the dopamine D_1- and D_2-receptors are differentially involved in this dual function.

Summary of the Hypothesis: Failure of Striatally Mediated Inhibition and Disinhibition of Thalamic Neurons Induces 'Positive' and 'Negative' Schizophrenic Symptoms, Respectively

The glutamatergic input from the cortex to the striatum appears to have two mutually antagonistic behavioral functions, i.e. one 'brake', mediated via the 'indirect' pathways, and one 'gas', operating via the 'direct' pathways. Failure of the 'break' and 'gas' would lead to positive and negative symptoms, respectively. Since these two components can be partly independent, it might explain why both positive and negative symptoms can exist side by side in varying proportions. Failure of glutamatergic pathways may lead to such mixed symptomatology. Such failure may, for example, be due to a genetically based defect of the NMDA receptor. Needless to say, the corticostriatal hypoglutamatergia would not necessarily be primary. It might be secondary to changes further upstream and related to the various structural and metabolic changes that have been repeatedly observed in the cerebral cortex and limbic-diencephalic system of schizophrenic patients [21–25]. Schizophrenic symptomatology can apparently also be induced by a downstream failure of striatal neurons, because such symptomatology may occur, and in fact precede motor disturbances, in case of Huntington's chorea [26].

If glutamate deficiency is causally involved in schizophrenia or in a group of schizophrenic patients, leading to a mixture of positive and negative symptoms, which would be mediated by the negative 'indirect' and positive 'direct' feedback loops, respectively, dopamine antagonists would be expected to improve the positive symptoms but, if anything, worsen the negative symptoms. Dopamine-releasing agents such as the amphetamines, methylphenidate etc. would be expected to worsen positive symptoms but to improve negative symptoms. This is in agreement with clinical experience [2].

Neuroleptics seem to be especially efficacious in paranoid forms of schizophrenia, and such forms are mimicked by dopaminergic agonists, acting directly on D_2- or D_1 + D_2-receptors or indirectly via dopamine release. In nonparanoid forms neuroleptics are less efficacious, especially if negative symptomatology prevails. These facts invite the speculation that paranoid schizophrenia is due to a hyperdopaminergic state, whereas non-paranoid forms may be due to a hypoglutamatergic state. One may also consider the existence of cases where an imbalance is induced by simultaneous hyperdopaminergia and hypoglutamatergia. Such cases should respond to neuroleptic treatment, though less completely than cases with pure hyperdopaminergia.

Future Outlook: Therapeutic Implications

The present discussion will be limited to the development of new pharmacological research tools/therapeutic agents. Just as modern schizophrenia research sprang out of drug research, pharmacological probes will in all likelihood continue to play a decisive role in the search for the causes of schizophrenia. Against the background of the hypothesis presented above, the most burning question deals with the glutamatergic system. It is reasonable to assume that PCP owes its psychotomimetic properties largely or entirely to its noncompetitive interference with NMDA receptor function and that other noncompetitive NMDA receptor antagonists are psychotomimetic. Whether competitive NMDA antagonists are also psychotomimetic, remains to be investigated.

Can glutamatergic agonists with antipsychotic properties be developed? A serious problem here is the potentially excitotoxic and convulsant properties of glutamatergic agonists. Can such effects be avoided, either by developing agonists acting at subtypes of glutamatergic receptors, which are not involved in excitotoxicity or cause convulsions, or by developing partial agonists, whose intrinsic activity is sufficient for antipsychotic activity but insufficient for inducing excitotoxicity or convulsions? Perhaps the development of agonists acting on allosteric sites of the NMDA receptors, such as the glycine and polyamine sites, might prove useful. Preliminary clinical trials with glycine and a glycine precursor have given somewhat contradictory results [27, 28]. The most important advantage of a glutamatergic agonist might be the ability to alleviate negative schizophrenic symptoms by improving the function of the positive feedback loop.

An interesting development is also underway in the dopamine receptor area. Clinical data on a D_1-receptor antagonist are as yet only preliminary. Given the interaction alluded to above, the relative importance of

D_1-receptors may be enhanced in a hypoglutamatergic state. If this is so, D_1 antagonists might turn out to be be very useful at least in a subgroup of schizophrenic patients. Conversely, the more or less selective D_2 antagonists which are now so much used, would be less useful by enhancing D_1-receptor stimulation via autoreceptor blockade and resultant activation of dopaminergic neurons.

An alternative way of achieving a balanced suppression of dopaminergic activity might be to use drugs acting by inhibiting dopamine release. Agents acting by preferentially stimulating dopamine autoreceptors would then be an interesting alternative. Several such agents, which seem to share the properties of partial dopamine receptor agonists, are now in clinical trials [29–33]. Since they are partial dopamine (D_2/D_3)-receptor agonists they have affinity also for postsynaptic dopamine receptors. With a sufficiently low intrinsic activity they will also be antagonistic on postsynaptic dopamine receptors, which will add to the antidopaminergic action. A suitable level of intrinsic activity might result in a satisfactory antipsychotic response, while at the same time extrapyramidal side effects would not be induced, because such effects require a high degree of postsynaptic dopamine receptor blockade.

Could dopamine agonists be used to counteract negative symptoms? Even though the amphetamines seem to possess the properties required for this purpose, they seem to be difficult to use in actual practice, in view of their too strong stimulant and addictive properties. Preferential dopamine autoreceptor antagonists may turn out to be useful for this purpose. They are mild stimulants, and excessive stimulation and addiction will presumably be avoided by the postsynaptic receptor blockade that shows up after higher doses. Examples of agents belonging to this class are the substituted aminotetralins (+)-UH232 and (+)-AJ76 [34]. Such agents are not yet available for clinical use but will hopefully be so within the next few years.

Currently the possible role of a number of dopamine receptor subtypes, recently discovered by means of cloning techniques, is intensely investigated [35]. For example, the D_4 subtype, belonging to the 'D_2-family', is attracting a great deal of interest, mainly because clozapine has been shown to exhibit a relatively strong preference for the D_4 versus the D_2 subtype. However, it is still an open question whether this preference will help to solve 'the clozapine engima'. Another member of the 'D_2-family' is the D_3 subtype. Also this subtype is attracting a great deal of attention, perhaps mainly because it is preferentially localized in the ventral, 'limbic' portions of the striatal complexes. This subtype has also been shown to be manufactured by the dopamine neuron itself, and thus presumably to serve as an autoreceptor, in addition to being a postsynaptic receptor. Certain dopamine receptor antagonists, previously found to act preferentially on dopamine autoreceptors, have been shown to possess preferential affinity for the D_3 versus the D_2 subtype. Examples of this group of agents are the previously mentioned (+)-AJ76 and (+)-UH232 [36]. Again, it cannot yet be decided if the preferential autoreceptor affinity of these agents is related to their preferential D_3 affinity. To solve the problem of the role of the dopaminergic receptor subtypes, it will be necessary to develop agents with sufficient selectivity for these various subtypes.

References

1 Domino E, Luby ED: Abnormal mental states induced by phencyclidine as a model of schizophrenia; in Cole JO, Freedman AM, Friedhoff AJ (eds): Psychopathology and Psychopharmacology. Baltimore, Johns Hopkins Univ. Press, 1973, pp 37–50.

2 Angrist B: Pharmacological models of schizophrenia; in Henn FA, Delisi LE (eds): Handbook of Schizophrenia, Vol 2. Neurochemistry and Neuropharmacology of Schizophrenia. Amsterdam, Elsevier, 1987, pp 391–424.

3 Lodge D, Aran JA, Church J, Davies SN, Martin D, Zeman S: Excitatory amino acids and phencyclidine drugs; in Hicks TP, Lodge D, McLennan H (eds): Excitatory Amino Acid Transmission. New York, Liss, 1987, pp 83–90.

4 Kornhuber HH, Kornhuber J, Kim JS, Kornhuber ME: Zur biochemischen Theorie der Schizophrenie. Nervenarzt 1984;55:602–606.

5 Kim JS, Kornhuber HH, Kornhuber J, Kornhuber ME: Glutamic acid and the dopamine hypothesis of schizophrenia; in Shagass C (ed): Biological Psychiatry. Amsterdam, Elsevier, 1985 pp 1109–1111.

6 Crow TJ: Two syndromes of schizophrenia as one pole of the continuum of psychosis: A concept of the nature of the pathogen and its genomic locus; in Henn FA, DeLisi LE (eds): Handbook of Schizophrenia, vol 2, The Neurochemistry and Neuropharmacology of Schizophrenia, New York, Elsevier, 1987, pp 17–48.

7 Carlsson M, Carlsson A: The NMDA antagonist MK-801 causes marked locomotor stimulation in monoamine-depleted mice. J Neural Transm 1989;75:221–226.

8 Carlsson M, Carlsson A: Dramatic synergism between MK-801 and clonidine with respect to locomotor stimulatory effect in monoamine-depleted mice. J Neural Transm 1989;77:65–71.

9 Carlsson M, Carlsson A: Interactions between glutamatergic and monoaminergic systems within the basal ganglia – implications for schizophrenia and Parkinson's disease. Trends Neurosci 1990;13:272–276.

10 Svensson A, Carlsson M: Injection of the competitive NMDA receptor antagonist AP-5 into the nucleus accumbens of monoamine-depleted mice induces pronounced locomotor stimulation. Neuropharmacology 1992;31:513–518.

11 Alheid G, Heimer L, Switzer RC: The basal ganglia; in Paxinos G (ed): The Human Nervous System. San Diego, Academic Press, 1990, pp 483–582.
12 Parent A, Paré D, Smith Y, Steriade M: Basal forebrain cholinergic and noncholinergic projections to the thalamus and brainstem in cats and monkeys. J Comp Neurol 1988;277:281–301.
13 Parent A, Lavoie B, Hazrati L-N, Côté P-Y: Chemical anatomy of the basal ganglia in normal and parkinsonian monkeys. 10th Int Symp Parkinson's Disease, Tokyo 1991, p 6.
14 Penney JB, Young AB: Striatal inhomogeneities and basal ganglia function. Movement Disorders 1986;1:3–15.
15 Alexander GE, Crutcher MD: Functional architecture of basal ganglia circuits: neural substrates of parallel processing. Trends Neurosci 1990;13:266–271.
16 Pan HS, Penney JB, Young AB: Gamma-aminobutyric acid and benzodiazepine receptor changes induced by unilateral 6-hydroxydopamine lesions of the median forebrain bundle. J Neurochem 1985;45:1396–1404.
17 Svensson A, Carlsson A, Carlsson ML: Differential locomotor interactions between dopamine D_1/D_2 receptor agonists and the NMDA antagonist dizocilpine in monoamine-depleted mice. J Neural Transm, in press.
18 Morelli M, Di Chiara G: MK-801 potentiates dopaminergic D1 but reduces D2 responses in the 6-hydroxydopamine model of Parkinson's disease. Eur J Pharmacol 1990;182:611–612.
19 Goodwin P, Starr BS, Starr MS: Motor responses to dopamine D1 and D2 agonists in the reserpine-treated mouse are affected differentially by the NMDA receptor antagonist MK 801. J Neural Transm (P-D Sect) 1992;4:15–26.
20 Svensson A, Carlsson ML, Carlsson A: Interaction between glutamatergic and dopaminergic tone in the nucleus accumbens of mice: Evidence for a dual glutamatergic function with respect to psychomotor control. J Neural Transm (Gen Sect) 1992;88: 235–240.
21 DeLisi LE: The use of positron emission tomography to image regional brain metabolism in schizophrenia and other psychiatric disorders: A review; in Nasrallah HA, Weinberger DR (eds): Handbook of Schizophrenia, vol 1. The Neurology of Schizophrenia, New York, Elsevier, 1986, pp 309–324.
22 Kirch DG, Weinberger DR: Anatomical neuropathology in schizophrenia: Postmortem findings, in Nasrallah HA, Weinberger DR (eds): Handbook of Schizophrenia, vol 1. The Neurology of Schizophrenia. New York, Elsevier, 1986; pp 325–348.
23 Buchsbaum M: The frontal lobes, basal ganglia, and temporal lobes as sites for schizophrenia. Schizophr Bull 1990;16:379–390.
24 Tamminga CA, Kaneda H, Buchanan R, Kirkpatrick B, Thaker GK, Yablonski MB, Holcomb HH: The limbic system in schizophrenia. Pharmacologic and metabolic evidence. Adv Neuropsychiatry Psychopharmacol 1991;1:99–109.
25 Tamminga CA, Thaker GK, Gao X-M, Shirakawa O, Kirkpatrick B, Alphs LD, Carpenter WT, Chase TN: Limbic system abnormalities in schizophrenia and neocortical alterations with deficit syndrome. Arch Gen Psychiatry, in press.
26 Mattsson B: Huntington's chorea in Sweden. II. Social and clinical data. Acta Psychiatr Scand 1974;(suppl 255):221–235.
27 Costa J, Khaled E, Sramek J, Bunney Jr W, Potkin SG: An open trial of glycine as an adjunct to neuroleptic in chronic treatment-refractory schizophrenics. J Clin Psychopharmacol 1990;10:71–72.

28 Tamminga CA, Cascella NG, Fakouhi TD, Herting RL: Enhancement of NMDA-mediated transmission in schizophrenia: Effects of milacemide: in Meltzer HY (ed): Novel Antipsychotic Medications. New York, Raven Press, in press.

29 Gründer G, Wetzel H, Hillert A, Benkert O: Roxindole, a dopamine autoreceptor agonist, in the treatment of negative symptoms of schizophrenia. Biol Psychiatry 1991;29:390S.

30 Klinke A, Klieser E: Antidepressant effects of the new dopamine autoreceptor agonist EMD 49980 (Roxindol) in acute schizophrenia. Biol Psychiatry 1991;29:410S.

31 Tamminga CA, Cascella NG, Lahti RA, Lindberg M, Carlsson A: Pharmacological properties of (–)-3PPP (preclamol) in man. J Neural Transm (Gen Sect) 1992;88:165–175.

32 Naber D: Efficacy and tolerability of the partial dopamine agonist SDZ HDC 912 in the treatment of schizophrenia. Biol Psychiatry 1991;29:409S.

33 Wiedemann K, Loycke A, Krieg JC, Holsboer F: EMD 49980 – a novel dopamine autoreceptor agonist in the treatment of schizophrenic patients with negative symptoms. Biol Psychiatry 1991;29:422S.

34 Svensson K, Johannson AM, Magnusson T, Carlsson A: (+)-AJ76 and (+)-UH 232: Central stimulants acting as preferential dopamine autoreceptor antagonists. Naunyn-Schmiedebergs Arch Pharmacol 1986;334:234–245.

35 Sibley DR, Monsma FJ Jr: Molecular biology of dopamine receptors. Trends Pharmacol Sci 1992;13:61–69.

36 Sokoloff PB, Giros B, Martres M-P, Bouthenet M-L, Schwartz J-C: Molecular cloning and characterization of a novel dopamine receptor (D-3) as a target for neuroleptics. Nature 1990;347:146–151.

Arvid Carlsson, Department of Pharmacology, University of Göteborg,
Medicinaregatan 7, S–413 90 Göteborg (Sweden)

Brunello N, Mendlewicz J, Racagni G (eds): New Generation of Antipsychotic Drugs: Novel Mechanisms of Action. Int Acad Biomed Drug Res. Basel, Karger, 1993, vol 4, pp 130–135

Dopamine (D_2)-Receptor Occupancy in Clozapine-Treated Patients as Measured by Positron Emission Tomography Using ^{18}FESP

J.W. Louwerens, J.A. Buddingh', S. Zijlstra, J. Pruim, J. Korf, A.M.J. Paans, W. Vaalburg, C.J. Slooff

Schizophrenia Rehabilitation Research Group (A collaborative effort of the Department of Psychiatry and PET-Center of the University of Groningen and Psychiatric Hospital 'Licht & Kracht', Assen), Groningen, The Netherlands

Introduction

The pharmacological properties of the antipsychotic drug clozapine raised questions about the concept that a high level of dopamine (D_2)-receptor blockade was required in order to achieve sufficient antipsychotic efficacy [1, 2]. This dopamine hypothesis was based on the fact that the neuroleptic drugs showed a very good correlation between the affinity for striatal D_2-receptors and their average clinically effective dose [3], while dopamine (D_2) agonists were psychotogenic in sensitive patients. This concept logically linked the antipsychotic efficacy of neuroleptics to their extrapyramidal side effects, which are caused by the drugs' interaction with dopamine neurotransmission as well. Clozapine was qualified as an 'atypical neuroleptic' because of the very low incidence of extrapyramidal symptoms related to its unique antipsychotic properties [4]. These clinical properties were in agreement with the relatively low affinity of clozapine for D_2-receptors [5], which was confirmed by the reports of some positron emission tomography (PET) studies that suggested low [6] to no affinity to D_2-receptors [7] in man as well.

In an earlier PET study we demonstrated the actually high occupancy of D_2-receptors by 'classical neuroleptics' in patients who were refractory to the therapeutic efficacy of these drugs [8]. Since then the treatment of most of our therapy-resistant schizophrenics has succesfully been changed to clozapine. This exceptional efficacy could not be explained by the combined D_2 and $5HT_2$ antagonism of clozapine [9]. This PET study was performed to determine the D_2 occupancy by clozapine in these formerly 'therapy-resistant' patients who all responded well to clozapine treatment.

Patients and Method

11 patients (7 males and 4 females, mean age 36, SD 8) with chronic schizophrenia, according to DSM III-r criteria (295.1, 295.3 and 295.9), all in (partial) remission on clozapine treatment, participated in this study. As controls 6 patients (5 males and 1 female, mean age 30, SD 10) of the same diagnostic groups, treated with classical neuroleptics, and 6 healthy volunteers (mean age 32, SD 7) were investigated according to the same design. Before PET scanning was carried out the treatment regimen in all patients was kept constant for at least 3 weeks. Informed consent was obtained from all patients and healthy volunteers. Clozapine was given in clinically effective dosages, ranging from 300 to 800 mg daily (mean 414 mg). Classical neuroleptics (fluphenazine, zuclopenthixol, flupenthixol and chlorprothixene) were administered in normal, and in 1 case even low, clinical dosages. All drugs were last administered 2 h before the patients received the intravenous bolus of the ligand ^{18}F-fluoro-ethyl-spiperone (18-FESP). PET scans, using 18-FESP as a ligand, were performed by means of a Siemens ECAT 951 PET scanner. 18-FESP was demonstrated to have comparable pharmacokinetic and pharmacodynamic properties as spiperone [10, 11]. This tracer shows an almost irreversible binding to D_2-receptors in vivo. During PET scanning the subjects were resting in supine position, without restrictions regarding the opening of the eyes and with ambient noise according to their own choice. Each patient received a rapid intravenous bolus of 2–5.5 mCi 18-FESP immediately prior to the first scan. 31 equally spaced, parallel planes, centered from the cantomeatal line, were scanned for 20 min. The subjects were carefully positioned using individual markers and laser beams. This procedure was repeated after 1, 2, 3, 4 and 5 h, though at 2, 3, 4 and 5 h after injection the scanning period was reduced to 15 min. Regions of interest (ROIs) were outlined individually on the reconstructed tomographic images, using preconstructed ROIs that were placed over the striatal areas, showing specific ligand binding. Unfortunately no MRIs could be obtained in order to determine anatomic relations. According to the localization of the striatal region showing the highest specific binding, 2 paramedial columns in the cerebellum (1 cm wide, 2 cm high) were determined in order to measure the aspecific ligand binding. All calculations on the scans were done in a comparable way to paramedial columns located in the occipital brain area after correction for attenuation. Three hours after tracer injection the highest activity is typically found in the dopamine-rich striatal areas, in contrast to the cerebellar hemispheres, where no specific binding takes place. The linear increase of the striatal to cerebellar ratio with time reflects the receptor binding by the treatment drugs. Therefore, either this ratio at a certain time after tracer injection or the slope of the ratio over time can be used as measure of D_2 dopamine receptor occupancy [12].

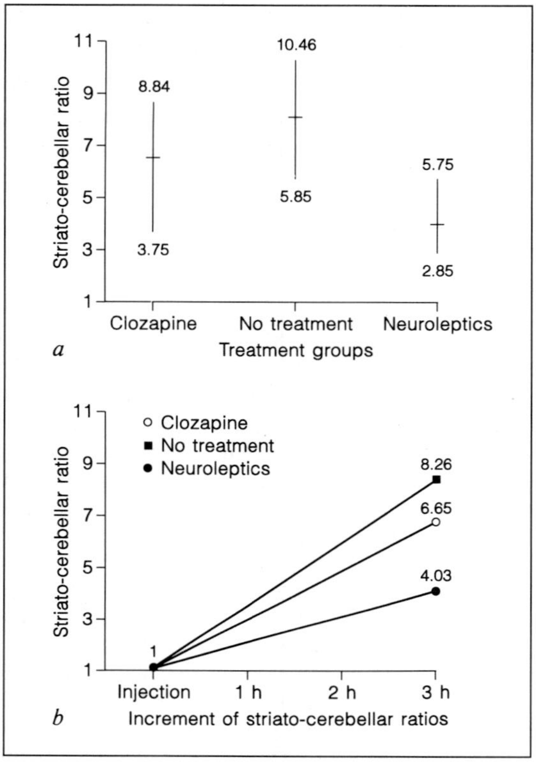

Fig. 1. 18-FESP binding 3 h after injection. *a* Striato-cerebellar ratios: different antipsychotic treatments. Values are means ± SD. *b* Slopes of striato-cerebellar ratios: different antipsychotic treatments.

Results

All patients treated with clozapine had detectable serum drug levels, ranging from 208 to 1,232 μg/l (average 596 μg/l). Only one of the patients treated with classical neuroleptics had an undetectable serum drug level (fluphenazine), the other 5 patients had low serum drug levels, but within the therapeutic range. All subjects showed an increase of the specific ligand binding after 3 h. The specific ligand binding was expressed as the striato-cerebellar ratio, using the striatal and cerebellar ROIs as described in the Method section. Figure 1a shows the means and ranges of the ratios of

18-FESP binding per treatment group, 3 h after tracer injection. In figure 1b the slopes of the increment of the average ratios per treatment group are indicated by lines connecting the ratios at the moment of tracer injection and the mean ratios 3 h later. Patients treated with classical neuroleptics did show a significantly ($p < 0.0002$) lower ligand binding in comparison to the healthy volunteers, reflecting an average occupancy of dopamine (D_2) binding sites by the treatment drugs of 58%. Clozapine caused a small decrease of ligand binding in comparison to healthy volunteers that did not reach significance (fig. 1a). Yet the difference of the slopes of the increment curves of clozapine-treated patients and healthy volunteers (fig. 1b) could represent an occupancy of dopamine D_2-binding sites by clozapine of 22%. An interesting finding was a difference between the treatment groups of specific 18-FESP binding in the occipital brain area. After 3 h the healthy volunteers and the patients treated with neuroleptics showed an average increase of the occipital cerebellar ratio to respectively 1.4 and 1.35. Yet in the clozapine-treated patients this ratio did not show such an increment (average 1.08). These differences were significant ($p > 0.01$).

Discussion and Conclusions

In an earlier study we demonstrated the extensive occupancy of dopamine D_2-receptors in the striatum by neuroleptics in patients who were resistant to the therapeutic effects of these drugs [8]. Clozapine treatment resulted in a definite improvement of the psychopathology of these patients. Yet, this clinical effect could not be attributed to an addition of a $5HT_2$ blockade to the dopamine antagonism, to which these patients did not respond [9]. This study shows that the patients who did respond to clozapine treatment have only a small number of dopamine (D_2) sites occupied. This in contrast to patients treated with classical neuroleptics whose dopamine (D_2)receptors in the striatum are occupied by the treatment drug to a large extent, comparable to the percentages of occupancy reported by other groups [12, 13].

The difference of specific 18-FESP binding in the occipital brain areas between neuroleptic-treated patients and healthy volunteers on one hand and clozapine-treated patients on the other hand suggests a specific binding of clozapine to receptors that have a clear affinity to spiperone in the occipital brain areas. This implies the presence of $5HT_2$-receptors in these areas, to which the neuroleptics showed a clearly lower affinity than clozapine.

$5HT_2$-receptors are present to a large extent in the putamen and caudate nucleus. The difference between the treatment groups of 18-FESP binding in the occipital brain areas as demonstrated here implicates the strong possibility that the calculated affinity of clozapine to D_2-receptors is an overestimation.

It can be concluded from these data that the clinical effect of clozapine is at most partially dependent on dopamine (D_2) antagonism, at least as far as this antagonism is of a type, to be traced using spiperone as a ligand.

Farde et al. [13] using C11-raclopride as a ligand did show a considerable dopamine (D_2) occupancy in the putamen in patients treated with clozapine 150–300 mg (40 and 65%). Our results indicate different dopamine (D_2)-binding characteristics for raclopride as compared to spiperone and probably other classical neuroleptics as well. Raclopride could give a comparable type of dopamine (D_2)receptor binding as caused by clozapine and possibly other atypical antipsychotic drugs. Spiperone on the other hand traces at most weakly the type of receptor binding caused by clozapine, while it interacts at the receptor site with classical neuroleptics in a comparable way to raclopride.

Hypothetically the binding sites for raclopride and spiperone at the dopamine rceptor are not identical. Raclopride binding could represent a subtype of the dopamine (D_2)-binding site. Clozapine could have a higher affinity to the raclopride-traced subtype of the receptor, that could be the major binding-site for 'atypical antipsychotics' at the dopamine (D_2)receptor. Spiperone on the other hand would, in this model, have a strong affinity to the total D_2-receptor-complex. About the clinical relevance of such a receptor subtype can be speculated.

References

1 Meltzer HY, Stahl SM: The dopamine hypothesis of schizophrenia: A review. Schizophr Bull 1976;2:19–76.

2 Meltzer HY: The mechanism of action of novel antipsychotic drugs. Schizophr Bull 1991;17:263–287.

3 Seeman P, Lee T: Antipsychotic drugs: Direct correlation between clinical potency and presynaptic action on dopamine neurons. Science 1975;188:1217–1219.

4 Kane J, Honigfeld G, Singer J, Meltzer HY: Clozapine for the treatment-resistant schizophrenic: A double-blind comparison versus chlorpromazine/benztropine. Arch Gen Psychiatry 1988;45:789–796.

5 Richelson E: Pharmacology of neuroleptics in use in the United States. J Clin Psychiatry 1985;46:8–14.
6 Farde L, Wiesel FA, Nordstrom AL, Sedvall G: D_1 and D_2-dopamine receptor occupancy during treatment with conventional and atypical neuroleptics. Psychopharmacology 1989;99:S28–S31.
7 Karbe H, Wienhard K, Hamacher K, Huber M, Herholz K, Coenen HH, Stöcklin G, Lövenich A, Heiss WD: Positron emission tomography with (^{18}F)methylspiperone demonstrates D_2 dopamine receptor binding differences of clozapine and haloperidol. J Neural Transm 1991;86:163–173.
8 Coppens HJ, Slooff CJ, Paans AMJ, Wiegman T, Vaalburg W, Korf J: High central D_2-dopamine receptor ocupancy as assessed with positron emission tomography in medication but therapy resistant schizophrenic patients. Biol Psychiatry 1992;29: 629–634.
9 Louwerens JW, Coppens HJ, Korf J, Slooff CJ, Paans AMJ, Vaalburg W: On the pathogenetic role of dopaminergic and serotonergic neurotransmission in long-term hospitalized psychotic patients. Schizophr Res 1990;3:52.
10 Zijlstra S, Louwerens JW, Buddingh' JA, Paans AMJ, Visser G, Slooff CJ, Vaalburg W, Korf J: The assessment of central D_2-dopamine receptor occupancy with positron emission tomography in long-term medicated schizophrenic patients; in Baron JC et al. (eds): Brain Dopaminergic Systems: Imaging with Positron Tomography. Dordrecht, Kluwer, 1991, vol 20, pp 181–189.
11 Wienhard K, Coenen HH, Pawlik G, Rudolf J, Laufer P, Jovkar S, Stöcklin G, Heiss WD. PET-studies of dopamine receptor distribution using (^{18}F)fluoroethylspiperone: Findings in disorders related to the dopaminergic system. J Neural Transm 1990;81:195–213.
12 Wolkin A, Brodie JD, Barouche F, Rotrosen J: Dopamine receptor occupancy and plasma haloperidol levels. Arch Gen Psychiatry 1989;46:482–483.
13 Farde L, Wiesel FA, Halldin C, Sedvall G: Central D_2-dopamine receptor occupancy in schizophrenic patients treated with antipsychotic drugs. Arch Gen Psychiatry 1988;45:71–76.

J.W. Louwerens, Schizophrenia Rehabilitation Research Group, PO Box 30001, NL–9700 RB Groningen (The Netherlands)

Brunello N, Mendlewicz J, Racagni G (eds): New Generation of Antipsychotic Drugs: Novel Mechanisms of Action. Int Acad Biomed Drug Res. Basel, Karger, 1993, vol 4, pp 136–142

Negative Symptomatology and the Therapeutic Use of Benzamides

Patrice Boyer, Yves Lecrubier

Inserm, Paris, France

Introduction

The so-called 'disinhibitory' [1] or 'energizing' neuroleptics [2] such as sulpiride, pimozide, pipotiazine and more recently amisulpride, can improve a cluster of negative symptoms (asthenia, avolition-apathy, psychomotor retardation, social withdrawal) very similar to those described by Andreasen and Olsen [3]. One of the important issues is then to determine if the main target of low doses of such neuroleptics is the subtype of schizophrenia itself or the negative syndrome per se. In order to partially answer this question two double-blind placebo-controlled studies using amisulpride were carried out in 104 and 125 patients, respectively.

First Study

Patients and Methods

The study was designed as a multicenter, double-blind placebo-controlled trial. Patients included had to fulfill Andreasen's criteria for negative schizophrenia, i.e. the presence of two at least of the following components with a severe intensity: affective blunting, alogia, anhedonia-asociality, avolition-apathy, attentional impairment. Patients also fulfilling criteria for positive schizophrenia were excluded from the study. Furthermore patients included had to have a score of at least 75 on the SANS (30 items) and a score of less than 60 on the SAPS. A 6-week washout period was mandatory for patients receiving neuroleptics or antidepressants (12 weeks in case of long-acting neuroleptic). Patients were randomly assigned to one of three groups: placebo, amisulpride 100 mg/day or amisulpride 300 mg/day. Treatment lasted 6 weeks during which all patients were hospitalized.

Table 1. Sex, age and diagnosis of patients in trial

	Total	Amisulpride		Placebo
		300 mg	100 mg	
	(n = 104)	(n = 36)	(n = 34)	(n = 34)
Sex M	65	22	20	23
F	39	14	14	11
Mean age, years	32.5	34.1	32.5	30.7
SEM	0.8	1.4	1.4	1.3
Range	19–49	19–45	20–49	19–45
DSM III subtype				
295.1X, disorganic	36 (34.6%)	14 (38.9%)	14 (41.2%)	8 (23.5%)
295.2X, catatonic	10 (9.6%)	4 (11.1%)	2 (5.9%)	4 (11.8%)
295.6X, residual	26 (25%)	4 (11.1%)	11 (32.3%)	11 (32.3%)
295.9X, undifferentiated	32 (30.8%)	14 (38.9%)	7 (20.6%)	11 (32.3%)

Results

A total of 104 patients were considered as evaluable (table 1). The three treatment groups were comparable concerning age, sex and DSM III diagnostic subtype at entry. The patients were rather young (mean 33 years). The mean duration of the illness was around 10 years with 59% of the patients having entered the disease with heavy deficit symptoms (mean duration of negative symptoms: 8 years). 51% of the patients were included after a 6- or 12-week washout period. The initial SANS scores were strictly comparable between the three treatment groups whatever the initial status (drug-free or treated) could be. In all three groups this initial score was high, about two-thirds of the maximal score (table 2). 19 patients dropped out during the treatment period: 6 due to exacerbation of positive symptomatology (4 in the placebo group, 1 in each amisulpride group), 4 due to worsening of the negative symptomatology (2 in the placebo group, 1 in each amisulpride group), 1 due to mixed deterioration (placebo group) and 8 for reasons independent of the symptomatology.

Efficacy

Groups were compared 2 by 2 according to Bonferroni's rule after comparison of the equality of variances. The main analysis was an analysis of completers (n = 84 data-based completers). The intent to treat analysis

Table 2. Illness and treatment history before inclusion

	Total	Amisulpride		Placebo
		300 mg	100 mg	
	(n = 97–104)	(n = 29–33)	(n = 29–33)	(n = 30–34)
Mean duration of illness, months	127 (9)	145 (17)	118 (16)	117 (16)
Range	4–336	4–306	12–336	4–326
Mean duration of negative symptoms, months	96 (9)	110 (18)	93 (13)	85 (14)
Range	2–329	2–293	6–329	2–282
Neuroleptic	49	18	15	16
Antidepressant	1	0	0	1
BZD	4	1	2	1
NL + AD	5	1	3	1
NL + BZD	27	12	6	9
AD + BZD	2	0	1	1
NL + AD + BZD	3	0	2	1
Never treated	13	4	5	4

Group totals vary because of missing data. NL = Neuroleptic; AD = antidepressant; BZD = benzodiazepine; standard deviations in parentheses.

was performed in a second step. All three groups showed a reduction in negative symptomatology after 6 weeks but a statistically important difference can be evidenced between the groups: mean total SANS score in the placebo group decreased only 22.8%, whereas in the amisulpride 100- and 300-mg groups it decreased 40.6 and 45.9%, respectively ($p < 0.022$). The final score in the placebo group (n = 74) barely drops below the 75 value, the minimum threshold for the inclusion of a patient. However, the final value in both amisulpride groups is close to 50 (table 3). The same is true for SANS subscores: reduction of scores is always more important in the two amisulpride groups than in the placebo group. These between-group differences are statistically significant for the subscales alogia ($p < 0.018$), avolition-apathy ($p < 0.036$) and attentional impairment ($p < 0.031$).

Table 3. Negative symptoms (SANS scores, n = 84, mean scores)

	D0	D42	Change, %	MANOVA
Total score				
AMI 300 mg	97.0	52.5	45.9	
AMI 100 mg	96.9	57.6	40.6	$p < 0.022$
Placebo	95.8	74.0	22.8	
Affective blunting				
AMI 300 mg	23.7	12.2	48.5	
AMI 100 mg	24.8	13.3	46.4	$p < 0.065$
Placebo	23.0	18.1	21.3	
Alogia				
AMI 300 mg	13.9	7.0	49.6	
AMI 100 mg	13.8	8.0	42.0	$p < 0.018$
Placebo	13.9	10.7	23.0	
Avolition/apathy				
AMI 300 mg	14.0	7.5	46.4	
AMI 100 mg	13.2	8.0	39.4	$p < 0.036$
Placebo	13.7	10.4	24.1	
Anhedonia/asociality				
AMI300 mg	19.4	11.4	41.2	
AMI 100 mg	19.1	13.1	31.4	NS
Placebo	18.3	14.6	20.2	
Attentional impairment				
AMI 300 mg	7.6	3.4	55.3	
AMI 100 mg	7.4	3.3	55.4	$p < 0.031$
Placebo	8.5	5.2	38.8	

Amisulpride (AMI) 300 mg: n = 30; AMI 100 mg: n = 30; placebo: n = 24.

They fail to reach significance for the subscales affective blunting ($p < 0.065$) and anhedonia-asociality. Intent to treat analysis shows the same differences at a more significant level, including the affective-blunting cluster. Positive symptomatology does not vary in a significant manner in all three groups. More important, there is a slight decrease of the SAPS scores in the two amisulpride groups, evidencing no productive-stimulant activity of low doses in this pathology. Extrapyramidal symptoms were rare at the beginning of the trial and did not change in a statistically significant way throughout the treatment period.

Second Study

Improvement in negative symptoms seems to be able to occur in schizophrenic patients with low doses of amisulpride. It appeared of interest to study the activity of the same compound for nonschizophrenic patients, for whom the existence of a negative syndrome represents a motive for consultation, because of the marked suffering or handicap it produces in social life.

Patients and Methods

A double-blind study compared amisulpride 25 and 50 mg to placebo during a 4-week period in 125 adult patients, 50 men and 75 women consulting a private psychiatrist for asthenia, loss of initiative and motivation, diminished performance and presenting therefore an isolated negative syndrome sufficiently marked to require a specific treatment. Patients were excluded if they fulfilled Feighner's primary depression criteria or if they presented a psychotic disorder, an anxiety disorder requiring anxiolytics or an asthenic state related to somatic conditions.

Statistical Analysis. Analysis of variance for quantitative variables after a test of equality of variances. Comparison of groups 2 by 2 according to Bonferroni's rule. Main analysis in intent to treat. Second analysis of completers.

Results

Dropouts. Twelve patients dropped out of the trial, 8 in the placebo group (inefficacy), 3 in the amisulpride 25-mg group (improvement) and 1 in the 50-mg group (refusal to continue the treatment).

Efficacy

The total score of the SANS (30 items) for the amisulpride 50-mg group differed from placebo on day 14 and 28 ($p < 0.001$) and from the 25-mg group on day 28 ($p < 0.05$). The 25-mg group differed from placebo on day 28 ($p < 0.05$). The evolution of the different components of the SANS is shown in table 4. A difference in favor of both the amisulpride 25- and 50-mg group versus placebo was also evidenced for the following

Table 4. Comparison of differences in improvement of 5 components of SANS

	Affective blunting		Anhedonia		Avolition		Alogia		Attention	
	D14	D28	D14	D28	D14	D28	D14	D28	D14	D28
50 mg > placebo	0.01	0.001	0.01	0.001	0.01	0.001	0.05	0.01	0.01	0.01
25 mg > placebo			0.05	0.01		0.05				
50 mg >25 mg							0.05	0.05		

items:general fatigue, difficulty in concentration, fall in libido, difficulty in falling asleep, headache, drowsiness, nocturnal and early morning awakening.

Discussion

As of 1980, Puech et al. [4] proposed a 'bipolar' hypothesis of dopaminergic activity in schizophrenia according to predominant symptomatology. If negative signs are related to a decrease in DA activity in the prefrontal cortex, the possible regression of these same signs under the action of DA agonists or low doses of neuroleptics is easily understandable. In the first study our findings show an important, statistically significant reduction of negative schizophrenic symptoms after a 6-week treatment with low doses of amisulpride compared to placebo. Reduction of scores is about twice as important (40–50%) in the amisulpride groups compared to placebo group (20–25%). Nevertheless the observation period was short (6 weeks), and there may be later onset of phenomena indicating reversion to nonresponse. In the second study the 'negative' symptoms of the nonpsychotic patients, as demonstrated by the SANS score, differ from that observed in schizophrenic patients by a low score for 'alogia' and 'attention disorders' and a moderate score for 'social withdrawal'. Although most patients of this study would be classified as DSM-III-R dysthymics, their syndromic homogeneity seems to be superior to their nosological one. Analysis in intention to treat and analysis of completers both show a significant improvement with low doses of amisulpride (50 and 25 mg). The responsivity of negative symptoms to amisulpride evidenced in both studies confirms the lack of specificity of this negative cluster [5]. In fact descriptive homogeneity does not imply causal uniformity. It might be possible that one part of the negative symptomatology described in schizophrenia is closely associated with this disease but that the general frame of this kind of pathology (encountered for example in dysthymic patients) is more closely linked to functional changes in neurotransmitter systems [6, 7].

References

1 Deniker P, Ginestet D: Les effets psychiques des neuroleptiques. Confront Psychiatr 1975;13:135–153.
2 Simon P, Lecrubier Y, Puech AJ: Classification des neuroleptiques. Rev Prat 1984; 34:589–594.
3 Andreasen NC, Olsen S: Negative vs positive schizophrenia: Definition and validation. Arch Gen Psychiatry 1982;39:789–794.
4 Puech AJ, Brochet D, Lecrubier Y, Simon P: Neuroleptiques: Activateurs ou inhibiteurs du système dopaminergique? Psychol Méd 1980;12:2427–2430.
5 Lecrubier Y: Multiple pharmacological mechanisms and clinical targets for neuroleptics: Should a more operational classification be considered? Psychiatr Psychobiol 1987;4:232–238.
6 Boyer P, Puech AJ: Determinants for clincial activity of neuroleptic drugs: Chemical substances, doses, assessments tools. Psychiatr Psychobiol 1987;4:296–305.
7 Boyer P, Lecrubier Y, Puech AJ: Treatment of positive and negative symptoms: Pharmacological approach; in Mod Probl Pharmacopsychiatry. Basel, Karger, 1990, vol 24, pp 152–174.

P. Boyer, INSERM, 100, rue de la Santé, F-75014 Paris (France)

Brunello N, Mendlewicz J, Racagni G (eds): New Generation of Antipsychotic Drugs: Novel Mechanisms of Action. Int Acad Biomed Drug Res. Basel, Karger, 1993, vol 4, pp 143–145

The Treatment of Schizophrenia: Cost- and Risk-Benefit Issues

Panel Discussion

Moderators: *Y. Lecrubier* (France), *H.Y. Meltzer* (USA)

Panelists: *A. Ament* (The Netherlands), *H. Beckmann* (FRG), *S.R. Hirsch* (UK), *C. Nemeroff* (USA), *J.M. van Ree* (The Netherlands)

Cost-Benefit Issues

Schizophrenia has a lifetime prevalence of about 1 % in the adult general population. The illness often occurs in adolescence or early adulthood and may then continue throughout adult life, either as chronic disabling disease or with acute exacerbations. The definition of the various clinical subtypes of schizophrenia deserves further nosological research and clarification and there is still a need for improving the classification of psychotic disorders in general and the schizophrenic disorders in particular.

The economic impact of schizophrenia poses major mental health problems to the community: it is estimated that about 2% of the total national product in industrialized countries accounts for the care of schizophrenic patients. In the United States there are about 2 million schizophrenic patients and in Western Europe, probably 3 million.

In the United States the direct costs of schizophrenia in 1985 were $20 billion and the indirect costs, namely attributable to loss of income and disability payments were about $25 billion. Thus the estimated direct and indirect cost of the illness to the US economy at the current time is about $55 billion. It is clear that the issue of benefit from new treatments will always have to be balanced against the risks.

A recent study conducted in the Netherlands, reported by Dr. Ament, showed that even if cardiovascular disease and cancer have the higher incidence, the major costs within the health care sector are associated with mental illnesses: about 20% of the total health care budget is used for mental diseases. From a more detailed study on the cost of schizophrenia it appears that 10% of the total costs for mental diseases is for the treatment of schizophrenia: thus in the Netherlands 2% of the health care budget is used for the treatment of schizophrenia. Analyzing further the cost categories, it emerges that the treatment of schizophrenia is mainly hospital-oriented: more than 80% of the costs is related to care in general psychiatric hospitals. Another interesting fact is about drug costs: drugs are only a minor element of the treatment of schizophrenia, less than 2% of the total costs, when compared to the treatment of other patients, which has an average drug costs of more than 10%.

The figures are somewhat different in the United States, where much more of the cost of health care comes out of private pockets and thus hospitalization has a higher cost. Approximately one third of Americans have no private health insurance. For those who have private health insurance, psychiatric coverage is the poorest. Thus after one or two hospitalizations it is usually totally exhausted. Patients at that point have to be turned over to indigent city or State hospitals or VA hospitals if they are fortunate.

Indirect costs of schizophrenia are also considerable and relate to higher mortality rate (partly due to suicide), lower productivity, absenteeism and early retirement. The lifetime risk of suicide in schizophrenia is around 7%, and it is even higher in treatment-resistant patients and the mortality rate, according to several studies, is twice as high as in the general population.

Risk-Benefit Issues

One of the most important issues in the long-term treatment with antipsychotics is the alarming increase in tardive diskynesia (TD) as a concomitant of neuroleptic medication. It has been reported at a number of international meetings that TD does not progress in everybody; it develops to a certain point and then remains more or less static. It has been reported that many patients do not find TD that disturbing, although some patients who are severely disfigured by TD do. Thus TD is no longer viewed as either irreversible or inevitably progressive, but often TD reaches the highest degree of severity within the first couple of months and it does not get any worse. Schizophrenics are actually almost oblivious to

TD and the symptoms do not bother them unless it causes them difficulty with breathing or swallowing. Moreover patients with TD could be isolated within their families and within the community: this is what makes the difference and which leads doctors to try to treat patients with atypical neuroleptics whenever it is felt that they may get TD. The social cost and the public health impact of schizophrenia cause a very heavy burden to society in general, but also to the family of schizophrenic patients who need appropriate assistance, since this is often a life-long illness requiring lifetime pharmacological and psychosocial treatments. Preliminary data show that family members of schizophrenics are at greater risk: they tend to develop certain psychiatric disorders, mainly depression, anxiety disorders, addictive and psychotic behavior more than relatives of patients with other diseases like somatic ones. In many cases, schizophrenics avoid receiving treatment, the patient is not treated, and the illness is not recognized by the family. The general practitioner does not recognize the illness because it has an insidious onset. People may live with various personality disorders for a long time and are considered a difficult person in the home. The risk of this illness is not only to the individual, but it is spread throughout the family and everyone that impacts around the family.

The development of new effective pharmacological management of schizophrenia still represents a major challenge for the next decade. Studies show that up to 25–50% of patients do not respond to classical neuroleptic treatment and that 35% of patients treated with neuroleptics will relapse each year. The failure may be due to the lack of efficacy of available treatments on the disease course, or to the high incidence of extrapyramidal (EPS) and other side effects and possibly to the lack of psychosocial support, which may also lead to poor compliance. Improved new treatments should be effective in controlling both positive and negative symptoms and should also demonstrate better tolerance. There is thus an unquestionable need for new and better antipsychotics and new compounds are being developed and explored in the treatment of schizophrenia in particular in neuroleptic-resistant patients, with less or possibly no EPS, including TD and improved clinical longterm prophylactic efficacy.

Subject Index